# THE MANAGEMENT OF TRANSSHIPMENT TERMINALS

## Decision Support for Terminal Operations in Finished Vehicle Supply Chains

# OPERATIONS RESEARCH/COMPUTER SCIENCE INTERFACES SERIES

Professor Ramesh Sharda
*Oklahoma State University*

Prof. Dr. Stefan Voß
*Universität Hamburg*

**Other published titles in the series:**

Greenberg / *A Computer-Assisted Analysis System for Mathematical Programming Models and Solutions: A User's Guide for ANALYZE*
Greenberg / *Modeling by Object-Driven Linear Elemental Relations: A Users Guide for MODLER*
Brown & Scherer / *Intelligent Scheduling Systems*
Nash & Sofer / *The Impact of Emerging Technologies on Computer Science & Operations Research*
Barth / *Logic-Based 0-1 Constraint Programming*
Jones / *Visualization and Optimization*
Barr, Helgason & Kennington / *Interfaces in Computer Science & Operations Research: Advances in Metaheuristics, Optimization, & Stochastic Modeling Technologies*
Ellacott, Mason & Anderson / *Mathematics of Neural Networks: Models, Algorithms & Applications*
Woodruff / *Advances in Computational & Stochastic Optimization, Logic Programming, and Heuristic Search*
Klein / *Scheduling of Resource-Constrained Projects*
Bierwirth / *Adaptive Search and the Management of Logistics Systems*
Laguna & González-Velarde / *Computing Tools for Modeling, Optimization and Simulation*
Stilman / *Linguistic Geometry: From Search to Construction*
Sakawa / *Genetic Algorithms and Fuzzy Multiobjective Optimization*
Ribeiro & Hansen / *Essays and Surveys in Metaheuristics*
Holsapple, Jacob & Rao / *Business Modelling: Multidisciplinary Approaches — Economics, Operational and Information Systems Perspectives*
Sleezer, Wentling & Cude/*Human Resource Development And Information Technology: Making Global Connections*
Voß & Woodruff / *Optimization Software Class Libraries*
Upadhyaya et al / *Mobile Computing: Implementing Pervasive Information and Communications Technologies*
Reeves & Rowe / *Genetic Algorithms—Principles and Perspectives: A Guide to GA Theory*
Bhargava & Ye / *Computational Modeling And Problem Solving In The Networked World: Interfaces in Computer Science & Operations Research*
Woodruff / *Network Interdiction And Stochastic Integer Programming*
nandalingam & Raghavan / *Telecommunications Network Design And Management*
guna & Martí / *Scatter Search: Methodology And Implementations In C*
avi/ *Simulation-Based Optimization: Parametric Optimization Techniques and Reinforcement Learning*
soukis & Mitra / *Decision Modelling And Information Systems: The Information Value Chain*
o / *Constraint And Integer Programming: Toward a Unified Methodology*
& Nuzzolo / *Schedule-Based Dynamic Transit Modeling: Theory and Applications*
Raghavan & Wasil / *The Next Wave in Computing, Optimization, And Decision Technologies*
Alidaee/ *Metaheuristics Optimization via Memory and Evolution: Tabu Search and Scatter h*
& Kuwahara/ *Simulation Approaches in Transportation Analysis: Recent Advances and ges*
obe & Yagiura/ *Metaheuristics: Progress as Real Problem Solvers*
Hartman/ *Graph Theory, Combinatorics, and Algorithms: Interdisciplinary Applications*
nandalingam/ *Telecommunications Planning: Innovations in Pricing, Network Design ment*

# THE MANAGEMENT OF TRANSSHIPMENT TERMINALS

## Decision Support for Terminal Operations in Finished Vehicle Supply Chains

DIRK CHRISTIAN MATTFELD
Technical University Braunschweig

Springer

Dirk Christian Mattfeld
University of Technology at Braunschweig
Germany

Library of Congress Control Number: 2005937009

ISBN-10: 0-387-30853-9 (HB) ISBN-10: 0-387-30854-7 (e-book)
ISBN-13: 978-0387-30853-1 (HB) ISBN-13: 978-0387-30854-8 (e-book)



Printed in the United States of America.

9 8 7 6 5 4 3 2 1

springer.com

# Contents

# Acknowledgments

This book was inspired by a three year lasting project with the goal to improve the efficiency of terminal operations at the port of Bremerhaven, Germany. This project was founded by Gerd Markus, the former State Secretary of the Senator for Ports, Transport and Foreign Trade (Ministry of the Federal State of Bremen).

The project was carried out by the teams of Prof. Dr. Herbert Kopfer, University of Bremen, and Prof. Dr. Carsten Boll, Institute of Shipping Economics and Logistics. As the project manager, I thank both of them. Furthermore, I would like to thank Dr. Weidong Zhang and everybody else who contributed to the project at the University of Bremen.

I would like to express my gratitude to the BLG Logistics Group team, particularly the managing director Bernd Kupke and the manager of terminal operations, Michael Reiter, for their continuous support. The project was supported by the BLG Logistics Group / Automobile Logistics GmbH & Co, and by the Program of Emphasis in Logistics at the University of Bremen.

Chapter 1

# INTRODUCTION

## 1.1 Freight Transshipment

We observe an ongoing trend towards globalized industrial production. Multinational companies aim at strategic competitive advantages by distributing their activities around the globe. As a result, the individual supply chains become longer and more complex. Next to the supply chain reliability, companies try to keep supply chains cost efficient and responsive, i.e. warrant short order fulfillment lead times (Sürie and Wagner, 2005). The above goals dictate low inventory levels at the stages of a supply chain as well as a high frequency of transports between the partners involved.

**Supply Chain Requirements.** Detailed performance measures for a supply chain are provided by the Supply Chain Operations Reference (SCOR) model (Supply-Chain Council, 2002). The SCOR model provides four levels with increasing detail of process modeling. In accordance to the process detail depicted SCOR metrics are defined for each level. Level 1 distinguishes metrics addressing the reliability of supply chains, their responsiveness, flexibility, cost and optionally their assets. On levels 2-4 these metrics are operationalized with respect to the process types source, make and deliver. Thus, as substantial activities of the deliver process, transport and transshipment are evaluated as an integral part of the supply chain.

The division of labor around the globe entails an increased volume of freight subject to worldwide transport networks (King, 1997). These networks are typically run by logistics service providers consigned with the transportation, transshipment and storage of goods. Service providers aim at economies of scale due to a consolidation of transport volume incurred from different customers (Tyan et al., 2003).

In order to warrant a high frequency of transports and a high utilization of the transport facilities at the same time, hub and spoke networks have emerged. Hub and spoke networks permit frequent main haul runs of high volume between the hubs of a network (Rodrigue, 1999). The spoke legs from the origin of goods to a sending hub and from the receiving hub to the destination of goods are accomplished by feeder traffic of low volume at a moderate frequency.

### 1.1.1 Reasons for Transshipment

The hub and spoke transportation policy requires the transshipment of goods at hub terminals. The transshipment will entail additional costs for handling and storage of goods. Nevertheless, transshipment can be either unavoidable or even advantageous for several reasons (Ballou, 1999, Chapter 8).

- The intermodal split demands a transshipment of goods. For goods feedered to a port, an intermediate storage is required in order to provide a safety margin for reliable loading and further shipment towards the final destination.

- Whenever the volume of goods to be shipped on a certain relation does not fully utilize a transport carrier, a bundling of goods of different origins at a transshipment hub is performed.

- At the receiving hub a break-bulk of goods may be required. Since retailers offer a great variety of products, a mixing of goods is performed at transshipment points.

- As a final reason for transshipment we consider the necessity of distribution centers holding a certain level of inventory in order to coordinate the stochastics in the supply and demand of goods.

Typically, there is no single reason for transshipment, but rather a combination of the above-described arguments legitimates the freight transportation via transshipment terminals (Mason et al., 2003). If, for instance, a modal shift is required at a receiving hub, the unavoidable transshipment can also contribute to a break-bulk of goods. In addition, the goods are temporarily stored at the terminal in order to provide a certain slack in the supply chain, before they are shipped to their final destination.

### 1.1.2 Terminal Operations

Although terminal operators are involved in a multitude of customer supply chains, central activities of transshipment can be identified. Terminal operations comprise the discharge of goods from the inbound carrier, the storage of goods for certain duration of stay, and the loading of goods on an outbound carrier. In some cases, a conditioning of goods applies by means of value added services. Finally, goods are retrieved from the storage facility and loaded onto a carrier for outbound transport. For an overview of container transshipment, see Vis and de Koster (2003); Steenken et al. (2004).

Dependent on the type of goods certain activities may or may not contribute significantly to the overall transportation, handling and storage costs. For bulk cargoes the distinction between unloading and storage can be negligible (Snyder and Ibrahim, 1996). For instance, grain is unloaded via suction and immediately blown into a silo for storage. To the contrary, in the container transshipment the distinct phases of unloading and storage are evident. Gantry cranes discharge containers to the quayside, before they are forwarded via straddle carrier to their storage location.

Independent of the type of goods to be transshipped, the productivity of terminal operations is a central figure of transshipment. The productivity denotes the volume of goods that is handled by a transport- or handling facility in a certain period of time. For instance a modern gantry crane is able to handle up to 80,000 twenty-foot equivalent units (TEU) per year. Alternatively, the production coefficient denotes the time span required in order to handle or transport a single commodity.

From an intra-firm's perspective, the productivity largely determines the variable costs of transshipment, thus, in the long run the mean productivity of operations is an important factor of competition for a transshipment terminal. From a customer point of view, the productivity of operations determines service times for carriers and forwarders and therefore can improve the attractiveness of a transshipment terminal (Fagerholt, 2000).

An improved productivity can also increase the flexibility and reliability of operations. Savings gained from an increasing productivity can be used either to react more flexible to customer demands or to provide a reliable stream of operations by reducing the system's nervousness. The former type of flexibility relates to sequential decision making without knowledge of the future, whereas the latter type of flexibility addresses situations where a given system is able to operate well in many different circumstances (Gupta and Somers, 1992).

Productivity improvements can be obtained by either investing in transport and handling facilities or by investing in the automated planning and scheduling support of terminal operations. Planning reduces the number of unfavorable decisions with respect to storage allocations, leading to costly relocations of goods. Obviously, relocations lower the overall productivity of operations and consequently compromise the competitiveness of a transshipment terminal.

## 1.2 Vehicle Transshipment

In this book we consider the support of terminal operations with regard to the transshipment of finished vehicles. Automotive supply chains have been intensively studied since long (Jones and Clark, 1990). As a result, intricate instruments for the performance measurement of suppliers have been developed by vehicle manufacturers (Schmitz and Platts, 2004). A detailed adjustment of production planning purposes between suppliers and manufacturers has already been achieved, such that the existence of further optimization potential is questionable (Gnoni et al., 2003).

### 1.2.1 Vehicle Supply Chains

The optimization potentials of vehicle delivery, however, have been neglected for long. Vehicles have either been produced in mass production for an anonymous market or they have been produced 'build-to-order' by accepting extremely long customer lead times. Nowadays, neither large stocks of finished vehicles nor lead times in the range of months are affordable. However, current vehicle supply systems are still not capable of supporting 'build-to-order' production concepts (Holweg et al., 2005). The intercontinental transport and transshipment system will play an important role for future vehicle distribution systems.

On the one hand, distribution has to be performed time effective such that customer expectations concerning the date of vehicle delivery are met. This requires a fine grained integration of third party transport and transshipment processes in the vehicle supply chain. On the other hand transport and transshipment activities have to be performed cost effective. This goal can only be pursued by an efficient management of transport and transshipment resources. Thus, the planning and scheduling of logistics resources may provide gains without compromising tough lead times entailed by supply chain management.

### 1.2.2 Transshipment Tasks

Globalized vehicle production has caused a dramatic overall increase of transportation volume in recent years. This development impacts the transshipment of finished vehicles particularly at European ports. Europe traditionally imports a vast number of vehicles, but has also strengthened its exports to oversea destinations in recent years.

Concerning import, vehicles arrive by carrier in large numbers and are either directly forwarded to the hinterland, or they are consolidated for ongoing carriage by feeder ship. As a third alternative, vehicles are stored at the compound of a distribution center located at the port. Typically distribution centers are run by the domestic subsidiaries of the manufacturing companies abroad. Concerning export, again large quantities of vehicles arrive via rail of feeder ship. Vehicles are unloaded and stay for the matter of consolidation before they are loaded for car-carrier transportation.

All of the above-mentioned reasons for transshipment apply, thus rather complex transshipment arrangements have to be carried out by the terminal operator. Since transshipments of vehicles are performed in a self-propelling fashion, the avoidance of damaging has to be pursued in the first case. Nevertheless, efficient operations have to be ensured, which seemingly contradicts the principle of safe and reliable operations. The anticipation of terminal operations by means of automated planning can enable the terminal management to provide safe and efficient operations.

Different to container and bulk-cargo transshipment, the terminal operations with regard to finished vehicles allow the definition of reasonably sized entities for planning. Charges of cars of similar type and destination are treated as entities, for which discharge, storage, retrieval, and loading activities can be anticipated. The balancing of terminal operations over time accounts for a safe and reliable treatment of vehicles. As for the transshipment of other cargo, the productivity of operations is a hallmark of competition also in vehicle logistics.

In this book a model for operations planning of transshipment tasks is developed. Furthermore heuristic algorithms for the automated planning support are investigated. This research benefits from insights gained from the involvement in a scientific project with the port of Bremerhaven. However, models and algorithms have been streamlined in order to outline their generic approach applicable to a wide array of transshipment issues.

## 1.3 Book's Overview

In Chapter 2 we take a look at the ongoing globalization process of the automobile industry. Thereby we explain the increasing transportation volume currently observed in the face of a stagnating automobile production.

We argue that an increase in number is caused by two impacts. First, production is outsourced in order to benefit from lower salaries in transition countries. Furthermore, currency exchange risks are alleviated. Finally, a production close to foreign markets may help to conquer foreign markets. We show, that next to the obvious reasons above, also the organizational structure of a multinational firm accounts to the vehicle transport volume imposed.

We describe challenges for outbound logistics resulting from distribution strategies in a globalized production before we describe the hub and spoke concept as an answer of the transport industry to the demand raised by modern vehicle distribution. Finally we sketch benefits gained by logistics service providers who aim at the holistic support of a distribution network.

In Chapter 3 we consider how ports do respond to the demand of the transport market by providing a suitable infrastructure and an organization which is capable to warrant a seamless modal shift in the supply chain.

Next to finished vehicles also used vehicle and trucks are transshipped. The transshipment of finished vehicles is further differentiated into short-sea and deep-sea transport requirements. To this end we differentiate segments of vehicle transshipment and describe their particular requirements concerning port infrastructure.

A significant portion of these vehicle flows is transshipped via ports of the so-called 'North Range', reaching from Zeebrügge in the south up to Hamburg in the north. We compare the business development for eight competing ports of this range, and discuss competitive forces attracting transshipment volume. It is observed, that the segmentation of the market in import and export as well as in different types of transshipment allows the competitors to occupy niches.

Regarding the commercially most attractive segments of the market a stiff competition exists, which can be influenced by the individual ports in two ways: First, strategic alliances with partners of the transportation chain enable complex logistic arrangements for the vehicle manufacturer. Second, the efficient, safe, and reliable handling of vehicles is a necessary condition for a long-term success.

In Chapter 4 we describe terminal operations to be performed in vehicle transshipment in terms of work processes. Particular attention is paid to the key-factors for efficient operations. We show, that next to the matter of a high productivity, the balancing of the operation effort is an important goal of terminal operations (Mattfeld and Kopfer, 2003). A balanced load ensures safe and reliable operations, a prerequisite for the transshipment of finished vehicles.

In Chapter 5 we formalize these considerations in a mathematical model integrating space allocation and personal deployment. Such a model is needed in order to assess the problem difficulty with respect to the number and type of variables, objective function and constraints. We discuss properties of the model developed and suggest a hierarchical problem separation of the integral model into two sub-problems.

In two subsequent chapters, we consider both problems in more generic terms. Therefore we propose straightforward models, which retain the general problem characteristics, but omit certain constraints of the real world problems at hand. For both problems we propose solution methods and perform computational investigations in order to verify the approaches.

In Chapter 6 we address the planning of transportation and storage capacity over time. Transshipment tasks compete for storage space in spatially distributed storage locations of finite capacity. Although the optimization model developed suggests considering the assignment of tasks individually, the Evolutionary Algorithm proposed evolves a period-oriented capacity utilization strategy. This capacity utilization strategy then controls a construction heuristic, which assigns tasks to periods and storage locations to tasks.

In Chapter 7 we consider the second sub-problem, namely the employment of personnel. We model this problem as a multi-mode task-scheduling problem with time windows and precedence constraints. We aim at determining a gang structure supporting reliable as well as efficient operations. For this end we propose a Tabu Search procedure that moves tasks between gangs. In order to determine the manpower demand of a gang we solve the corresponding one-machine scheduling problem by an iterated Schrage heuristic.

In Chapter 8 we discuss the integration of operations planning into an existing IT-infrastructure of a transshipment hub. By example of the vehicle transshipment hub Bremerhaven we outline general concepts of integrating planning issues as a re-engineering activity. We start with a description of commonly used IT-functions supporting the execution of terminal operations, before we sketch the interfaces used in order to synchronize the information system level and the execution level.

We then focus on software modules for a user interface to planning. Finally, we discuss details of the integration of planning by viewing the planning activity as a business process. We handle this issue on the level of the requirement definition, the design specification and the implementation description. We conclude this final chapter with a discussion of the impact of planning for the vehicle transshipment hub of Bremerhaven.

Chapter 2

# AUTOMOBILE PRODUCTION AND DISTRIBUTION

**Abstract** In this chapter we take a look at the ongoing globalization process of the automobile industry. Thereby we explain the increasing transportation volume currently observed in the face of a stagnating automobile production.

We argue that an increase in number is caused by two impacts. First, production is outsourced in order to benefit from lower salaries in transition countries. Furthermore, currency exchange risks are alleviated. Finally, a production close to foreign markets may help to conquer foreign markets. We show, that next to the obvious reasons above, also the organizational structure of a multinational firm accounts to the vehicle transport volume imposed.

We describe challenges for outbound logistics resulting from distribution strategies in a globalized production before we describe the hub and spoke concept as an answer of the transport industry to the demand raised by modern vehicle distribution. Finally we sketch benefits gained by logistics service providers who aim at the holistic support of a distribution network.

## 2.1 Trends in Automobile Production

Automobile production has always been an important segment of the industries of North America, Japan and Western Europe. The technical complexity of automobiles led at an early stage to pioneering production techniques, which later on were often transferred into other industrial sectors. In this way automobile production was a driving force in the development of the industrial nations in the 20th century.

### 2.1.1 A Quick Glimpse to History

From the beginning of the automobile industry, manufacturers have recognized the importance of the economies of scales resulting from the

*Table 2.1.* Distribution of the share (in %) of foreign direct investments undertaken by the German automobile industry. Three years are selected to typify the last two decades. Investments in regions of particular interest are displayed in bold face.

| Region | 1981 | 1990 | 1998 |
|---|---|---|---|
| EU central | 15.0 | **23.1** | 14.2 |
| EU periphery[a] | 10.8 | **24.4** | 7.7 |
| America | **40.4** | 23.3 | 12.7 |
| Africa[b] | **10.6** | 5.8 | 1.3 |
| Asia[c] | — | 1.0 | 0.4 |
| Other industrial countries | 21.4 | 20.0 | **52.7** |
| Other transition countries[d] | — | 2.4 | **11.0** |

[a] Ireland, Portugal and Spain.
[b] Rep. of South Africa and Nigeria.
[c] excluding China.
[d] including China.

high investments required for vehicle mass production. Before the Second World War, Ford and General Motors aimed at extending their market share by undertaking investments preferably in Europe. At that time, the increasing automobile market allowed a co-existence with the domestic manufacturers of Europe.

After the Second World War, Japan successfully entered the market, mainly because of a significant reduction of the production costs achieved by means of new production and distribution systems (Amasaka, 2002). The Japanese manufacturers produced automobiles in extremely large numbers and shipped these vehicles to their export markets worldwide, see Borstnar (1999) for a comprehensive treatment of this topic.

During the 1980s, Japanese manufacturers began to produce in closer proximity to their export markets abroad, while American and European manufacturers started to engage Japanese production technology. Although the relative advantage of the Japanese manufacturers gradually declined, Japan had already established itself as a third important region of automobile production.

The economic pressure triggered by the Japanese competitors has led to foreign investments also being made by European manufacturers. Table 2.1 lists direct foreign investments of German manufacturers for selected years as compiled by Spatz and Nunnenkamp (2002). One clearly observes the massive engagement in South- and Central America, as well as in South Africa in the 1980s.

In the 1990s, fallen customs barriers in Europe have caused a shift of the investments abroad towards the low-income countries at the European Union (EU) periphery. Recently, the political developments in the countries of Central and Eastern Europe have drawn foreign direct investments towards this emerging region (Tulder and Ruigrok, 1998).

The continuous investment of the automobile industry has led to a change in the share of the production abroad compared to the domestic production volume. In 1990 already 26% of the German cars were being produced abroad; this figure had increased to almost 44% by 2000 (Verband der Automobilindustrie, 2002). Since this trend also holds for manufacturers located in other industrial countries, we are going to take a closer look at the way investments are undertaken.

### 2.1.2 Shift Towards Emerging Markets

Since is manpower-intensive, one can observe a drift of production towards low-income countries. However, the complexity of operations requires a relatively high industrial standard, such that transition countries like South Africa or Indonesia receive particular attention as candidates for automobile production (Pontrandolfo and Okogbaa, 1999). As transition countries are also emerging markets for automobile manufacturers, local distribution and services can easily be established as a side effect of an investment in production facilities.

Some transition countries like Malaysia protect their local production (Proton) by imposing import tariffs, sometimes in the range of a multiple of the product's value. According to Lüders (2001), Malaysia levies up to 300% customs duty on automobile imports. Custom rates can usually be alleviated or even be avoided by adding a local content to import automobiles. For the case of an import market of limited capacity, cars are typically shipped in parts, which are then assembled by making use of the local workforce.

Dependent on the level of pre-assembly of the parts shipped, we distinguish between "semi knocked down" (SKD) and "completely knocked down" (CKD). With SKD, a relatively small portion of cars are delivered in a semi-assembled fashion, whereas with CKD only parts are delivered for later assembly in the destination country. In the same vein, finished vehicles are sometimes referred to as "completely built up" (CBU). SKD/CKD solutions do not require an intricate management of supply and inventory of parts, which eases the implementation of an assembly plant as an initial investment in a transition country.

Whenever an emerging market exceeds a certain capacity, the country is considered for a "transplant", abbreviating a transferred plant. Besides the supply to the local market, automobiles are also produced for

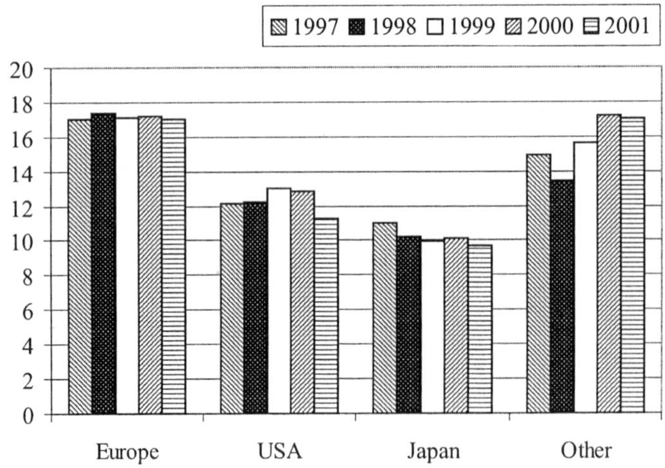


*Figure 2.1.* Automobile production in millions between 1997 and 2001 and main regions of production. The figures are compiled by the author on the basis of material provided by Ward's Communication (various issues).

export. In this way, economies of scale can be achieved and, moreover, the export surplus gained can be balanced with incidental import customs (Kuhn, 1998). As an example, Kuhn describes the development of a South African plant from a CBU importer to a SKD producer, and finally to a CBU exporter of niche products, i.e. right steering versions of certain Mercedes models.

Typically, local suppliers contribute to the local content of a transplant, leading to new sources of supply for other locations of the manufacturer, too. In this way, an interwoven logistical network emerges. For a comprehensive study with regard to the development of the Bayrische Motorenwerke (BMW) production at Spartanburg, South Carolina, see Dornier et al. (1998, Chapter 9-1).

Apart from the above-described foreign investments, local investments in automobile production exist, too. Borstnar (1999) describes the role of the Korean government as entrepreneur for the automobile industry. Similarly, China is expanding the production of automobiles to serve its protected local market, while currently lacking international competitiveness.

Summarizing, we observe a gradual decline of automobile production in the triad, namely North America, Japan and Western Europe, in favor of other countries, as illustrated in Figure 2.1. With the exception of China, these countries at least partially produce for export, while the main markets for automobiles continue to remain in the industrial

| principles of hierarchy | degree of hierarchical control | |
|---|---|---|
| | weak | strong |
| internationalization | multidomestic | worldwide |
| globalization | multiregional | transregional |

*Figure 2.2.* Possible configurations of multinationalization.

countries of the triad. Therefore, we can expect an increase of vehicle transportation due to the diversification of automobile production.

### 2.1.3 Multinational Configurations

Transportation demand may be estimated in more detail by considering the imbalances between production and sales within a geographical region on a per model level. For instance, a surplus of Mercedes M-class production in Tuscaloosa, USA, cannot offset the existing demand for S-class cars in North America, so that S-class cars have to be shipped from Sindelfingen in Germany.

In order to assess whether demand imbalances contribute to the transportation demand in general, and whether they will increase in the future, global strategies of the automobile industry are considered in the following. See Pontrandolfo and Okogbaa (1999), for an introduction into global manufacturing issues, Bélis-Bergouignan et al. (2000), for a transition model of global strategies for the automobile industry and finally Kuhn (1998) for the implementation of the model at Daimler Benz AG (now Daimler/Chrysler).

With respect to the above literature, Figure 2.2 considers different configurations of multinational corporations with respect to the principles of hierarchy and the degree of hierarchical control. Internationalization as a principle of hierarchy refers to the process of expanding a corporation's sphere of operation without any change to its initial structure. This means that only unilateral flows exist, originated at the center. In contrast to this, globalization involves multilateral flows in a poly-centric system. Both principles may appear with either a weak or a strong hierarchical control of the firm's activities.

- A *worldwide* configuration refers to an automobile manufacturer building and selling more or less identical automobiles worldwide, governed by a strong central control. Next to sales, also marketing, after-sales and service activities are centrally controlled. Although this configuration may not fully satisfy foreign customers, it allows the manufacturer to switch easily between production in different geographical regions and transportation into these regions. Many Far Eastern manufacturers like Toyota stand proxy for the worldwide configuration.

- A *multidomestic* configuration keeps the limited product range of the worldwide configuration, but adds geographical differentiation with respect to service activities. National subsidiaries may appear as domestic firms, although their products are not. The advantage of greater adaptation is paid by a weak control in the multidomestic configuration. Volkswagen may serve as an appropriate example of this configuration. Transportation will gradually decline because imbalances of volume between different regions are no longer resolved by a central control.

- In a *multiregional* configuration, products are designed and built with respect to the consumer taste of a certain geographical region. Products tend to become incompatible with each other and cannot be exchanged between different regions anymore. As an example, consider the configuration of General Motors, e.g. Vauxhall (UK), Opel (Germany), and Holden (Australia). The domestic companies have been virtually independent, as they act on regional markets with only weak central control. Consequently, the transport volume is only marginal.

- The *transregional* configuration adds stronger control to the multiregional configuration. It respects the differentiation of consumer taste with regard to geographical regions, however it aims at unifying product lines wherever possible (e.g. Daimler Chrysler). Regional satellites rather than domestic subsidiaries act under a strong central coordination of activities (Cohen and Mallik, 1997). Generally, the transregional configuration has obvious advantages as it combines customer orientation and production efficiency. Furthermore, it can make use of locally available resources and warrants a firm's flexibility.

The trend towards stronger hierarchical control schemes goes hand in hand with an increasing transportation volume, particularly for the transregional configuration. With regard to economies of scale, typically the worldwide production of a vehicle model is awarded to one single plant only. This increases the distance moved per automobile produced, which in turn will lead to a substantial increase in vehicle transportation (Cullen, 1998).

The additional costs encountered for vehicle transportation are limited, if not negligible. Kiedel (2001) names 250 to 500 Euro for an intercontinental transport. In discussions at a workshop of the German Operations Research Society (GOR) on vehicle delivery in 2003, a range of 2-3% of a vehicle's value is amounted for transportation purposes. Thus, an increase of transportation distance will not necessarily level advantages gained from economies of scale in production.

In order to adopt this hypothesis of increasing transportation volume, evidence for the transregional configuration of the automobile industry is needed.

Kobrin (1991) suggests an *Index of Transnational Integration* in order to measure globalization at an industry level. The index is defined as the proportion of international sales of a firm that are intrafirm. The logic behind this measure is that global supply chain coordination will lead to increased intrafirm flows of value. Highly integrated industries are motor vehicles (0.43), followed by electronic components (0.38). A less integrated industry is, for instance, glass production (0.11). Although Kobrin has performed this investigation on 1982 data of the USA, we can state a mainly transregional configuration for the automobile industry.

Cohen and Mallik (1997) repeated this investigation for 1989 data and achieved comparable indices (motor vehicles (0.39), electronic components (0.43), glass products (0.20)). Interestingly, the index decreased for highly integrated industries and increased for non-globalized producing industries. The authors argue that the index does not take care of outsourcing, which has become a hallmark of globalization today. Thus, the automobile industry remains a highly integrated industry and continues its course of integration through significant outsourcing activities.

## 2.2 Challenges for Outbound Logistics

Literature outlines the ongoing convergence of consumer taste resulting in the design of so-called 'world cars'. However, model classes like for example 'van' or 'utility' are still designed for the US market. As a new development of the transregional configuration, vehicles designed for a specific market are available also on other markets. Therefore as an obstacle, a manufacturer's globalization strategy may lead to a large variety of car models which may be far from a unified product line.

Although consumers can choose from an almost confusing variety of vehicle models, they are not willing to wait for the delivery of the vehicle for too long. Miemczyk and Holweg (2001) argue that although most customers would rather wait instead of buying an incorrectly equipped vehicle, the majority of North American consumers are not willing to wait more than three weeks. Similar observations have also been made for the UK market, where 61% of customers want their vehicle to be delivered within 14 days or less.

In order to cope with customer expectations, today on average 55, 70, and 20 days of sales are held in Europe, USA, and Japan respectively (Miemczyk and Holweg, 2001). The exceptionally small storage level for Japan results from the practice of satisfying domestic orders directly from stock, which is originally intended for export. According to the same authors, in Europe savings of 10 billion Euros have been estimated for the elimination of finished vehicle stocks. In order to cut inventory-holding costs, manufacturers have set up "built to order" programs, aiming at a time span of approximately 14 days from order to delivery. Recently, only 3-day car production spans have been pronounced (Holweg and Miemczyk, 2003).

Outbound logistics of finished vehicles are confronted with a set of constraints, which are hard to meet simultaneously:

- The transportation volume will increase due to foreign investments of manufacturers and furthermore due to their shift towards transregional configurations.

- A vast variety of vehicle models exists from which customers may choose. As a result the number of items per model sold on a regional market will be relatively small.

- In order to avoid extensive buffer stocks, the transportation lead-time has to be sufficiently short. This requires a fast transportation mode and, probably more important, a high frequency of transports.

- Since a vehicle model is typically produced in one plant worldwide, the transportation distance to the customer market can be very great.

Even major manufacturers will ship volumes, which does not justify the exclusive utilization of transport facilities. Hence, vehicle manufacturers have to accept sharing a liner service in order to warrant a high frequency of transports at reasonable costs (Hines et al., 2002). For this reason a worldwide transportation network has been evolved which is subject of the following considerations.

### 2.2.1 Distribution Strategies

From the viewpoint of the vehicle manufacturer, the annual volume of a transport relation is fragmented into frequent consignments of relatively small vehicle quantities. The duration of transport has to be reasonably short in order to implement "built to order" concepts. Thus, from the manufacturers' point of view, frequent direct shipments are desirable, but can be hardly afforded (Hines et al., 2002).

We identify four logistics solutions to the problem of cutting down the delivery times on regional markets. All of them have been already implemented, although not every solution fits the needs of every manufacturer's distribution policy.

- Vehicles are "built to stock" and a considerable inventory of vehicles is held at the import compound from which customer orders are supplied.
- Vehicles are "built to stock", but the inventory level is reduced by providing a final customer-order driven assembly step at the import compound.
- Vehicles are "built to order", and a point-to-point line haul service with frequent intermediate port calls is engaged.
- Vehicles are "built to order", and an efficient trunk haul is engaged with the drawback of additional transshipments and short-sea feedering.

Strategic developments of ports have to fit the above distribution policies of vehicle manufacturers in order to survive in the long run. In the sequel we describe hub and spoke approaches, which are slowly winning recognition in order to implement competitive distribution policies.

### 2.2.2 Hub and Spoke Distribution

Supply lines are becoming more fragmented and finished vehicles have to be shipped in smaller quantities. This causes a steady increase in the demand for more port calls to be made by the car carriers for a given volume of vehicles. The greater number of port calls also increases the travel time, which contrasts with the expectations of the manufacturers for "built to order" vehicle delivery.

Carrier companies are interested in making fewer port calls while shipping a number of vehicles. A remedy is offered by a hub and spoke-like design of the delivery network, as is already common for container transportation. Trunk haul routes between hub ports are serviced without intermediate port calls. As an alternative to road transportation, relatively small and flexible feeder ships service the legs from the vehicle plant to the sending hub, as well as from the receiving hub to the final destination.

The trend towards the feedering of legs in vehicle transportation has not become widely accepted so far, chiefly because of the reluctance of the manufacturers to accept transshipment related to the very high damage levels that were common in car shipping (Drewry, 1999). Nevertheless, a trend towards the transshipment of vehicles is already evident. Manufacturers will accept the general need for the transshipment of cars because this will enable the carriers to reduce transportation times and increase the frequency of port calls.

In 1978, the United States' Airline Deregulation Act' led to the reorganization of flight legs in 'hub and spoke' systems operated by the major airlines. This change of structure has been accompanied by a multitude of publications concerned with 'hub and spoke' systems. Next to the issue of optimal hub locations (O'Kelly, 1986) the effect on the utilization and frequency of flight connections has been subject to research (Phillips, 1987). The German reader is referred to the comprehensive PhD. thesis of Mayer (2001). Several advantages can be outlined:

- The *economies of density* refer to the consolidation of flows of goods. The improved utilization of transportation capacity can be used to improve the frequency of transport.

- The *economies of scope* refer to composite advantages due to the utilization of shared facilities. Hubs are used by a multitude of logistic relations, such that fixed costs can be shared among all relations involved.

- The *economies of scale* refer to the increased volume of goods for hub-to-hub relations. Among other digression effects on the costs, savings due to larger transportation and transshipment facilities exist.

However, also the disadvantages of hub and spoke transportation systems have to be taken into account:

- The *need for transshipment* may cause inconvenience or may lead to a general reluctance of customers. The time required for transshipment, however, may be over-compensated by the higher frequency of transports offered.

- A *congestion of traffic* has been observed for hubs, leading to a capacity overload of the transshipment facilities. As a common wisdom, an overloaded system entails a deteriorating efficiency of operations.

Although a hub and spoke network for vehicle transportation can provide benefits for manufacturers and for carriers at the same time, the additional transshipment of vehicles remains as a strong argument against the constitution of a hub and spoke system. As a remedy, hub feedering can also be performed by means of the hinterland transportation.

A lengthening of the hinterland transport may be accepted in order to achieve transshipment at a port with a high frequency of carrier callings.

Depending on the distribution policy pursued, manufacturers will consider the need for rapid transport, and in this event costs and transportation times of an additional feeder shipment have to be traded of against a hinterland detour (Abrahamsson et al., 1998). With respect to Western Europe, currently noteworthy short-sea feedering is performed to Scandinavia, UK and to the Iberian Peninsula only. For these routes, land transportation is applicable at extremely high costs.

Hinterland detouring for the matter of accessing a port with a high frequency of car carrier calling has led to a redefinition of the term "hub" in the context of vehicle transports. "Although it is common to speak of hub and spoke ports in the call shipping business, the term here is used to describe the volume of cars moved, rather than denoting the number of feeder movements" (Drewry, 1999).

### 2.2.3 Logistic System Leadership

Up to now the demand of nearly all manufacturers has been just for a basic transport service — and not for an integrated series of functions or operational management capabilities. However, actors in the vehicle logistics market anticipate a significant increase of contract logistics during the next few years. For instance, Drewry (1999) reports that the Wallenius Wilhelmsen Lines' objective is "that logistics should represent up to 30% of the group's revenue within 5-10 years, compared to only 2% today".

Indeed, Wallenius Wilhelmsen Lines has taken over the "Richard Lawson Autologistics Group" and it has started to run a terminal in Zeebrügge under its own management. Although the port has attracted significant transshipment volume in recent years, this development does not necessarily depend on the increasing vertical integration of Wallenius Wilhelmsen Lines. Temporary alliances between different partners of the logistic chain have long been known in order to harmonize tenders with respect to manufacturer demands.

One important reason for the reluctance of manufacturers to get involved with dedicated transshipment arrangements may stem from their dreaded loss of independence. The distribution of vehicles is a vital function in the environment of globalized production and manufacturers will not hand over the responsibility to a third party (Holocher, 2000). However, the demand for "built to order" will lead to complex vehicle distribution policies, which may also accelerate the transfer from basic transport to complex logistic arrangements.

## 2.3 Summary

We have discussed the ongoing globalization of automobile production. We have shown that the transportation volume of finished vehicles increases due to production abroad in general, and due to the transregional configuration of automobile manufacturers, in particular. Since there are good reasons for automobile manufacturers to spread production even further, we expect a continuous growth of vehicle transports for the future.

We have described several ways of organizing the worldwide distribution of vehicles and we have shed light on the possible role of ports in this process. "Built to order" concepts are likely to prevail, which will stress the importance of a suitable distribution policy in years to come. Ports are playing a particular role as they are acting as a hub within a complex transport arrangement. Besides the seamless integration of intermodal transshipment in the supply chain, ports may position themselves as third-party logistics provider controlling the entire supply chain of an automobile manufacturer.

In the next chapter, we are going to discuss prerequisites for a competitive port. However, competitiveness of a port is not restricted to the transshipment of finished vehicles. Also used cars and trucks as well as other heavy machines are transshipped in ports. Next to the requirements concerning port infrastructure the logistic demand differs for short-sea and deep-sea transports. By comparing the development of 8 major ports of the Lowlands, i.e. Belgium, The Netherlands and Germany we explain how ports respond to the demand caused by globalized production.

# Chapter 3

# INTERMODAL VEHICLE TRANSSHIPMENT

**Abstract**

Due to almost exhausted capabilities of shipping technology, emphasis has shifted from transportation to transshipment issues in order to accelerate distribution processes even further(Rodrigue, 1999). Ports respond to this demand of the transport market by providing a suitable infrastructure as well as an organization capable to warrant a seamless modal shift in the supply chain.

Next to finished vehicles also used vehicle and trucks are transshipped. The transshipment of finished vehicles is further differentiated into short-sea and deep-sea transport requirements. To this end we differentiate segments of vehicle transshipment and describe their particular requirements concerning port infrastructure and suprastructure, where particular attention is paid to the latter term referring to assets like available knowledge or research capital (Nijkamp and Ubbels, 2004).

A significant portion of these vehicle flows is transshipped via ports of the so-called 'North Range', reaching from Zeebrügge in the south up to Hamburg in the north. We compare the business development for eight competing ports of this range, and discuss competitive forces attracting transshipment volume. It is observed, that the segmentation of the market in import and export as well as in different types of transshipment allows the competitors to occupy niches.

Regarding the commercially most attractive segments of the market a stiff competition exists, which can be influenced by the individual ports in two ways: First, strategic alliances with partners of the transportation chain enable complex logistic arrangements for the vehicle manufacturer. Second, the efficient, safe, and reliable handling of vehicles is a necessary condition for a long-term success.

## 3.1 Vehicle Transport Market

As a consequence of the logistic challenges faced by automobile manufacturers, a logistics third-party market for the transport, transshipment and storage of finished vehicles has been evolved (Tyan et al., 2003). The distribution of CBU vehicles can be differentiated by the transportation distance and the vehicles to be shipped.

Typically, deep-sea and short-sea transportation are distinguished, where short-sea transports compete with rail and road transportation. Next to the transportation of new cars, also trucks and second-hand cars are shipped, each of them with special logistical requirements. In the following, we briefly describe the segments of vehicle transportations and the service provider performing the transport.

### 3.1.1 Deep-Sea Segment

When the intercontinental transport of significant number of vehicles started after the Second World War, ships built for bulk cargo were used. The loading and unloading was done by crane operations, which came along with high damage rates despite a low overall operations productivity. During the 1960s, cargo vessels were equipped with flexible decks for vehicle storage and lashing facilities for safe transportation. Such vessels had a capacity up to 2000 vehicles. In order to provide a sufficient loading productivity, these early car carriers were equipped with up to six cranes. Each crane was able to load 30 vehicles per hour, such that a total of 180 vehicles per hour could be handled during port calls (Kiedel, 2001).

Roll on / Roll off (RO/RO) transport was pioneered for the Scandinavian transport market during the 1920s but has not been taken up later on. Only in 1957 a RO/RO vessel with sideways mounted loading ramps was built, but again, this principle of transportation was not pursued.

With the rise of Japanese production, efficient transportation was needed in order to ship a huge number of vehicles to their markets abroad. This demand has led to the development of specially designed car carriers with a capacity of up to 6,200 vehicles sailing at approximately 20 knots, c.f. Figure 3.1. Pure Car Carriers (PCC) are distinguished from Pure Car and Truck Carriers (PCTC), which can be utilized more flexibly at the expense of a smaller capacity.

The car carrier market is distinct from most other shipping markets. Its determining characteristic is the power of the vehicle manufacturers. In this competitive sector, shipping contracts are awarded on a strategic/tactical basis lasting for 3-5 years on average. The supply of car carrying space increased by 17% between 1995 and 1999 and perhaps

*Figure 3.1.* Construction scheme of a modern car carrier with a capacity of approximately 6000 vehicles. The vessel contains 13 decks of approximately 1.7 meters height. To provide space for trucks, some of the decks can be lifted hydraulically. A ramp system comparable to a multi-storey park deck makes all decks accesssible in a self-propellent fashion (©Wallenius Wilhelmsen Line).

thirty-fold since 1970. In 1999 the fleet of car carriers amounted to 380 ships. Until now, the fleet has grown by an additional 110 car-carriers (Marle, 2003).

Only six carrier companies share 90% of the deep-sea market (Woodbridge, 2001). Above all NYK from Japan and the Wallenius-Wilhelmsen Lines from Scandinavia play major roles in this field. The business changed from main haul runs from Japan to Europe and North America, which accounted for over 65% of all deep-sea vehicle movements in 1985, to a liner service structure with fixed day calls at specific ports. For instance, NYK's Europe Connect service is operated on a twice-weekly schedule from Antwerp, Rotterdam, Bremerhaven and Southampton to Port Klang, Singapore, Keelung and Tokyo.

Another scheduled line is described in Kiedel (2001): The Wallenius-Wilhelmsen Lines starts from Göteborg, Sweden, leaves Europe at the 5th day of its journey after calls at Bremerhaven, Zeebrügge and Southamp-

*Table 3.1.* Estimated worldwide intercontinental vehicle transportation in 2001 for factory-new cars, light commercial vehicles and second-hand vehicles in thousands.

| Origin | | | | Destination |
|---|---|---|---|---|
| North America | Europe | Asia | Total | |
| — | 790 | 2300 | 3090 | North America |
| 75 | 80 | 330 | 485 | South America |
| 210 | — | 1900 | 2110 | Europe |
| 0 | 70 | 400 | 470 | Middle East |
| 60 | 150 | 120 | 330 | Africa |
| 70 | 300 | — | 370 | Asia |
| 0 | 60 | 435 | 495 | Australia |
| 415 | 1450 | 5485 | 7350 | Total |

ton. On the 13th day of its journey New York is the first stop at the American east coast. Further stops are Charleston, Jacksonville and Brunswick. On the 27th day of its journey it has already passed the Panama-Channel and calls at Mazatlan on the American west coast. After Stops in Port Hueneme and Tacoma the journey proceeds in the direction of Japan, where the line ends at the 33rd day of the journey. In Japan, the vessel takes up another scheduled line back to Europe again.

Estimates of the total deep-sea transportation volume differ between 7.5 and 8 million vehicles annually. Table 3.1 shows a conservative estimate of the main vehicle flows according to a 2002 presentation of HUAL, another important carrier. The figures have been verified by the author with respect to statistics presented by Auto & Truck International (2002). It is clearly observed that vehicle flows from Asia to Europe and North America dominate. However, due to the relatively high export rates to North America, the import and export flows to and from Europe are balanced.

Carrier companies aim at a high capacity utilization of their car-carriers due to balanced in- and outbound load while making calls at only a few ports (Mourão et al., 2001). Their lines have to follow the shift of vehicle production, recently towards the Philippines, Thailand, Mexico, Argentina, Australia and Brazil, which has led to a further increase of transportation volume, although carrier lines are becoming even more fragmented (Marle, 2003).

### 3.1.2 Short-Sea Segment

The short-sea segment covers water-ways within continental distances. Transports can be distinguished into direct shipments and hub feedering between large, deep-sea oriented terminals and smaller ports at remote locations. Typically, the short-sea segment is serviced by vessels up to 100 meters length with a capacity of less than 1000 vehicles (Kiedel, 2001).

Short sea shipping resembles other shipping markets to a greater degree. The shipping companies concerned have been much smaller and the market much more fragmented (Cullen, 1998). In the following we focus on the transportation volume arising between countries of Western Europe. Companies active within the finished vehicle distribution markets include Axial (Europe), AutoLogic (UK, Benelux), Cobelfret (Europe), Gefco (Europe), Groupe CAT (Europe), E.H. Harms (Europe), STVA (Europe), Transportvoiture (Belgium), Stifa, Hödlmayr, Horst Mosolf, ATG Autotransportlogistik GmbH, HN Autotransport, Transfesa and Zust Ambrosetti (MarketLine, 1998).

Since the Channel Tunnel between France and the UK as well as the Oeresund Brigde between Denmark and Sweden have been built, short-sea transport competes directly with rail and truck in Europe. For distances shorter than 500 kilometers truck transportation is advantageous due to its greater flexibility. Rail is an appropriate mode of transportation especially when the number of vehicles transported fully occupies a block-train of 600 meters length. With respect to Germany, block trains are routed e.g. from Munich to Bremerhaven within 24 hours, whereas smaller quantities of vehicles will take up to 48 hours to cover the same distance (Kiedel, 2001).

Unfortunately, all vehicles using rail transport have to be carefully washed in order to avoid permanent damage. Also flying sparks caused by overhead contact lines of the electric locomotives can harm the vehicle's finishing. Therefore, German Rail and DaimlerChrysler currently are pioneering a closed railway carriage called "tube" in order to avoid damages due to dirt, sparkles and vandalism (Bahntech, 2004). Whenever feasible, sea transportation is a cheap, safe and reliable alternative to land transportation.

Figure 3.2 shows the export volumes originating from six important vehicle producer countries into various regions of Western Europe compiled by the author from Auto & Truck International (2002). Significant volumes are exported from France and Germany to many other West European countries, whereas particularly France, Germany, Spain and the UK import significant quantities from almost all producer countries.

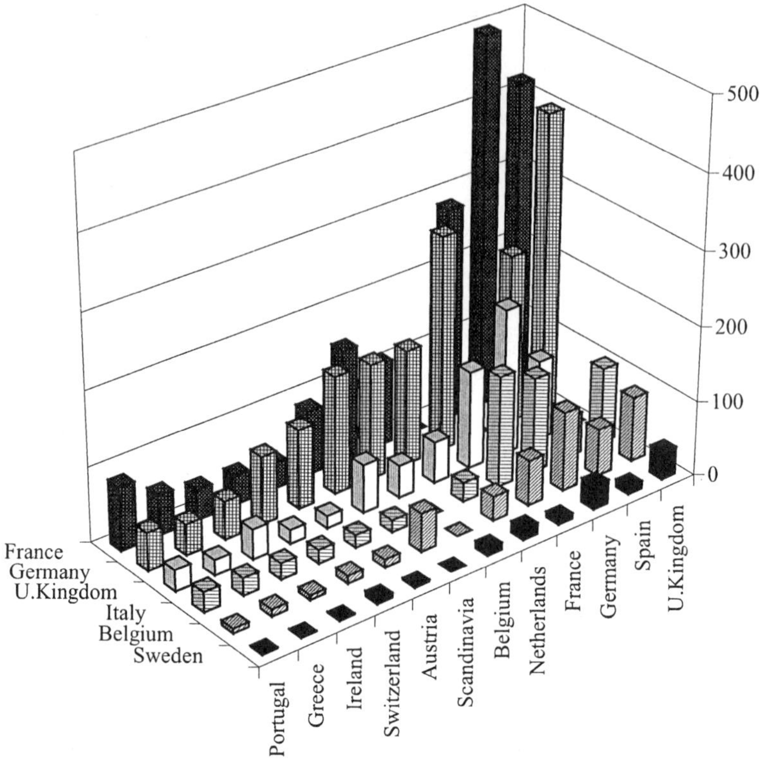


*Figure 3.2.* Volume of inner-Europe vehicle transportation in 2000 of CBU vehicles in thousands. One axis depicts the six European countries with substantial vehicle production, whereas the other axis depicts the inbound vehicle volumes for countries of Western Europe.

In the case of the UK imports amount to over one million vehicles, for which short-sea transportation is the first choice.

Another important short-sea market is the connection between Germany and Scandinavia. Here, Volvo cars are received from Sweden for transshipment to other European countries. On the other hand, significant volumes of UK, German and French vehicles are exported to Scandinavian countries. Short-sea transportation benefits from a general reluctance of deep-sea carriers to visit Baltic waters.

### 3.1.3 Second-Hand Automobile Segment

From a European perspective, second-hand car transportation is clearly an export market, because people prefer to drive factory-new cars in countries of high income. Consequently, older cars are exported to regions of lower income. Regions of destination are East Europe and the Middle East for relatively high valued cars and West Africa for lower valued cars, often to be repaired. South America and South East Asia are typically supplied by the USA and Japan.

From West Europe, approximately 600,000 automobiles are exported by sea-freight per year. Most of these vehicles are destined for West Africa; the port of Cotonou, Benin, alone discharges 250,000 vehicles per year. A significant portion with destination Africa or the Middle East shown in Table 3.1 is considered to be second-hand cars. Next to PCC/PCTC shipping it is common practice that cargo vessels arriving from West Africa load vehicles as back haul.

Although second-hand car transportation does not seem to be very attractive for logistic service providers, one should consider that the power of second-hand vehicle traders is less than the extremely strong negotiation power of vehicle manufacturers and consequently could be more profitable. Obviously, second-hand vehicle transportation is a niche segment, which does not necessarily imply that profits are low.

### 3.1.4 High and Heavy Segment

Another niche market is the so-called high and heavy segment dealing with the import and export of trucks, buses, construction and agricultural vehicles. Next to self-propelling vehicles, heavy goods like machines or large boats are placed on special trailers, which can be moved inside the PCTC this way. At European ports a total of 155,000 items with approximately 1.5 million tons are conveyed per year. Since the dimension of goods differ, high and heavy goods are usually quantified in tons or in Car Equivalent Units (1 CEU = $10m^3$).

A specially sealed area suited for heavy loads has to be provided on the quayside. However, the increasing transportation rates in this sector may be worth the investments to be taken. For the transport PCTC are needed, which are equipped with variable decks allowing a maximal height of 6 meters. Furthermore, armed loading ramps of 100 tons capacity are needed (usually 20-30 tons).

According to Marle (2003), the carrier company NYK is tooling up 10 newly built PCTC with such equipment in order to address this increasing market. In the past, many high and heavy goods have been shipped as assembly kits in containers. Nowadays, vehicles are fully assembled already at the originating port and transshipped via RO/RO trailers onto PCTC. This strategy avoids additional assembly effort at the destination country.

## 3.2 Competition Factors for Ports

Ports have to respond to the requirements of the automobile manufacturer's distribution policies. As we have seen, these requirements with respect to the transport distance to be covered as well as the type of vehicle to be transshipped. However, competition factors can be determined which contribute to the success of an individual port (Beykal, 2005).

### 3.2.1 Competitive Port Infrastructure

In order to assess a port's suitability for vehicle transshipment, we may consider its *accessibility*, its *extensibility* (including its current stage of extension), and finally its *facilities* installed.

**Accessibility.** Recently, Herfort (2002a) has claimed, that it is the hinterland distance to the production plants of a manufacturer which primarily determines the disposition to utilize a certain port, and is therefore critical to the development of a port. Unfortunately, the location of a port is immovable and not subject to management decisions.

Also the accessibility of the berthing facilities and their distances to the open sea is of concern. However, ports located directly at the coast often suffer from inconvenient transport connections to the hinterland. Rail and highway connections are essential; access for barges via rivers or channels is advantageous in the competition for transshipment volume.

Also the tide is of concern. Modern PCC or PCTC are equipped with sideways mounted loading ramps. Operations require a maximal blade angle of 12 degree in order to avoid damages of the vehicle's underbody or spoiler. Therefore, in regions of high tide, sluices are needed to divide tidal waters from waters of steady height suitable for port operations (Kiedel, 2001).

**Extensibility.** Another key-advantage is the disposability of storage space. Cullen (1998) has reported that "Bristol for example has storage capacity for 100,000 vehicles. The total value of this buffer stock, when it is full, is over 1.7 billion Euros." Since Far Eastern imports have declined during recent years, the availability of storage space is getting less competitive.

However, the quality of storage facilities is still important. Next to sealed ground, for instance, Bremerhaven has built multi-story car parks of 500,000 square meters in order to provide sheltered storage. Kuhr (2000) outlines that multi-story facilities allow operations with a very high productivity, because their location near the berthing facilities reduces the vehicle access times.

**Facilities.** Requirements with respect to facilities are threefold. First, added-value services are to be offered, second operations shall comply with quality management and third, IT-systems warrant a seamless integration of accompanying data flows.

- The assembly of optional vehicle equipment is referred to as added-value services. Generally, import transshipment requires a damage inspection, a de-waxing of vehicle surfaces and a pre-delivery inspection. Typically, polishing and anti-corrosion treatment complete pre-delivery services. Furthermore, ports offer customization and body conversions as well as fitting of parts or accessories and quality repairs. Finally, some ports also provide the reconditioning of second-hand cars.

- In order to warrant safe and reliable operations, quality management in accordance with the ISO 9000 standard is enforced by the manufacturers, who are primarily concerned about vehicle damages. Across the industry, the damage levels have dropped consistently to currently level off between 0.5% and 1.0%. Anything over this range is likely to result in the loss of contracts from the manufacturers, i.e. a switch of transshipment volume to other competing ports (Drewry, 1999).

- As a final prerequisite an information system with data links to customer IT-systems is demanded (Hesse, 2002). The use of an IT infrastructure suitable for tracking and tracing of vehicles cannot be over-estimated with regard to future competitiveness.

In the following, we aim at an empirical verification of the state of progression of terminal transshipments. First, we introduce ports belonging to the strategic group of the so-called North Range. We describe their geographical location and their properties, before we summarize the development of their inbound and outbound transshipment volume during the last decade. In the following we differentiate individual ports by their market share, which allows us to draw conclusions about the emergence of hub and spoke structures. Finally the consideration of the transshipment with respect to the vehicle segments described reveals the existence of dominating ports within each segment.

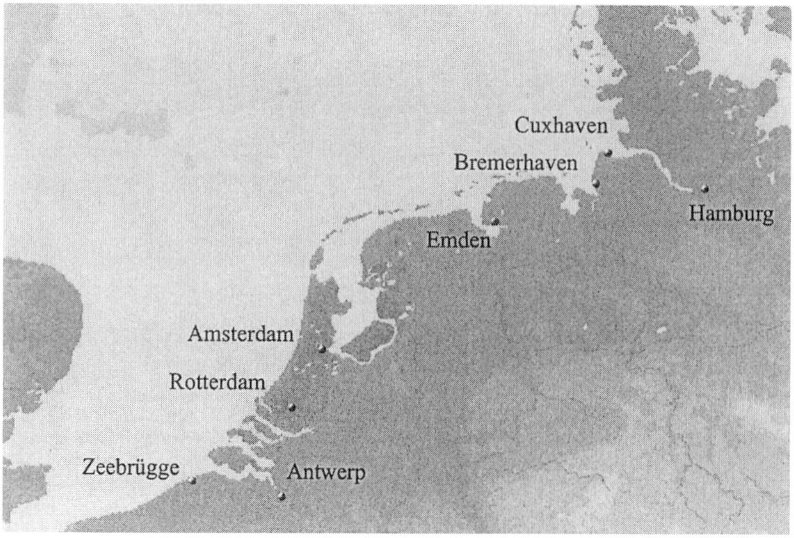


*Figure 3.3.* Competing vehicle transshipment ports in the Zeebrügge—Hamburg North Range.

## 3.2.2 Ports of the North Range

Since trade statistics subsume automobile transshipments under the category RO/RO cargo, precise statistics are not readily available. We confine ourselves to the investigation of eight ports in the Zeebrügge—Hamburg range, termed the North Range, for which reliable data is available. These ports compete for transshipment volume, because they are located in geographical vicinity. For an overview of the location of the ports considered see Figure 3.3.

The ports in Belgium are Zeebrügge, located closest to the UK, Antwerp at the Rhine, and in the Netherlands Rotterdam and Amsterdam. Additionally, Vlissingen and Ghent contribute with relatively small vehicle transshipment volumes, but these ports have been omitted due to lack of reliable data. In Germany, Emden is considered at the River Ems as well as Bremerhaven and Cuxhaven at the estuaries of the rivers Weser and Elbe. Finally, Hamburg is taken into account as an inland port. Together, these ports tranship 60% of all European deep-sea vehicle transportation. The other 40% are distributed to Scandinavian, Mediterranean, and UK ports.

For the empirical investigation three sources have been consulted. A long-term development of transshipment volume with respect to the ports considered is available for inbound and outbound separately from

*Table 3.2.* Characterization of ports of the Zeebrügge—Hamburg range. The symbols '–', '0', and '+' denote a fairly negative, neutral, or positive setting of the capabilities.

| port | accessibility | | | | facilities | extensibility |
|---|---|---|---|---|---|---|
| | sea | truck | rail | barge | | |
| Zeebrügge | + | 0 | 0 | – | + | + |
| Antwerp | – | 0 | 0 | + | + | + |
| Rotterdam | 0 | + | + | + | 0 | 0 |
| Amsterdam | 0 | + | + | + | 0 | 0 |
| Emden | 0 | 0 | + | + | 0 | + |
| Bremerhaven | + | + | + | 0 | + | 0 |
| Cuxhaven | + | + | 0 | 0 | – | 0 |
| Hamburg | – | + | + | + | – | – |

Holocher (2000), compare Figures 3.4, 3.5 and 3.6. For the year 2001, the author has conducted an empirical study which differentiates the transshipment volume by market segments (Lieske, 2002). Furthermore, a detailed assessment of the transshipment facilities of all of the above-mentioned ports has been carried out. Recently, Bremenports (2003) has completed a market study of the competitive situation of the Bremerhaven port, which largely supports figures provided by Lieske and Holocher.

Although, in principle, the proximity of ports would allow a substitution of transshipment volume, their capabilities differ somewhat, see Table 3.2. Zeebrügge, as a newly established port possesses excellent facilities with a great potential for expansion. The port is located at the coast, providing fast access for car carriers. Its hinterland connections, however, are currently very limited. Also Antwerp possesses extensible facilities, but its seaward access is adverse because of its inland location.

Rotterdam and Amsterdam are well known for container and bulk cargo transshipment respectively with excellent hinterland connections. The terminal facilities of both ports are currently devoted to other trades. Emden has a long history of exporting Volkswagen (VW) vehicles and has good rail links. Unusually, VW holds a share in the port's transshipment company.

Bremerhaven is Germany's largest vehicle port with an excellent infrastructure and connectivity. However, its extensibility appears to be limited. On the other hand, after a decade of low investment, and despite limited capabilities, Cuxhaven currently attracts short-sea transports to Scandinavia. Hamburg with its dominance in container transshipment

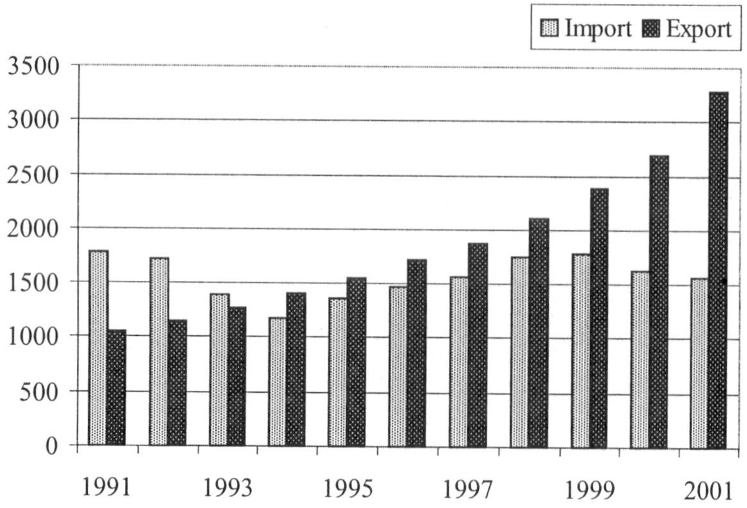


*Figure 3.4.* Vehicles per year transshipped inbound and outbound via important ports of the North Range between 1991 – 2001 (in thousands).

does not possess a viable infrastructure for vehicle transshipment. Summarizing, no obvious advantage with respect to the transshipment volume can be derived from the infrastructure overview.

### 3.2.3 Development of Transshipment Volume

The development of vehicle import and export differs for the ports considered. While inbound transshipments have remained almost stable over the last decade, outbound transshipments have increased by a factor of 3 since 1991 (see Figure 3.4). A gradual decline of inbound transshipment is caused by the current deterioration of Far Eastern exports. The increase of the outbound transshipment volume can be explained by the success of German manufacturers. German exports have increased by 30%, from 2.5 million up to 3.6 million vehicles per year during the 1990s.

However, the strong increase of outbound transshipments may also indicate an increasing hub and spoke share. Hub and spoke transport is typically counted twice at the receiving hub, because ongoing carriage by feeder ship is accounted as an additional outbound volume. Therefore the increase of volume may also be due to a rise of feeder transports to and from hubs. Since we can expect feeder transports to many secondary ports outside the North Range, their additional inbound volume will not be considered. Merely the outbound volume of hubs in the North

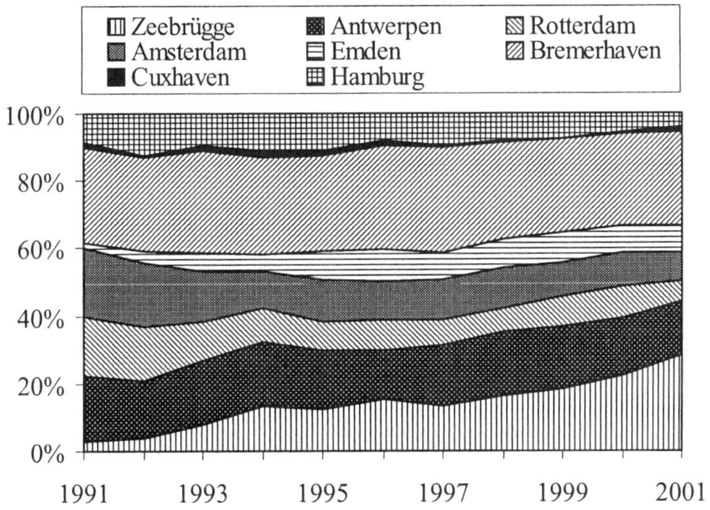


*Figure 3.5.* Development of the percentage of vehicle inbound volume transshipped via eight ports of the North Range between 1991 and 2001.

Range may have contributed to the above figure. In order to investigate the effect of feedering, we consider the development of inbound and outbound volumes separately in the following.

Figure 3.5 shows the market share of the inbound transshipment volume with respect to the ports considered. During the last decade a significant change of market share has taken place. For the ports in the Low Countries the volume has been shifted from Rotterdam and Amsterdam towards Zeebrügge. Antwerp's share has remained almost constant. Emden has increased its percentage of inbound transshipment due to Volkswagen imports from Spain and Mexico. Hamburg's share has continuously declined, whereas the large share of Bremerhaven has remained constant.

Interestingly, Zeebrügge, as the port with the bottom quality of hinterland connection, has shown an increase of the inbound market share of 20%. Obviously, its location at the western end of the North Range is giving it an important role in inbound transshipment. Carriers arriving from the Far East favor Zeebrügge because of the mileage savings compared to other ports of the North Range. Moreover, Zeebrügge's relatively short distance to Spain and its direct proximity to the UK may also have contributed to its dominant role. Apart from Cuxhaven ev-

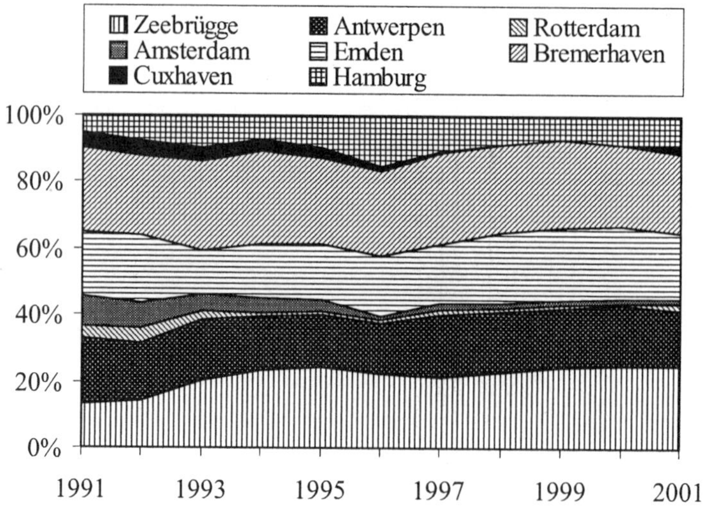


*Figure 3.6.* Development of the percentage of vehicle outbound volume transshipped via eight ports of the Zeebrügge—Hamburg North Range between 1991 and 2001.

ery port can rely on import compounds from at least one manufacturer, making its inbound market shares are relatively evenly apportioned.

The corresponding figures for outbound transshipments are shown in Figure 3.6. Here, Cuxhaven, Amsterdam and Rotterdam have been almost driven out of the market (although recently Cuxhaven started to export VW to Scandinavia again) (Harman, 2000). Zeebrügge has slightly increased its share, while Antwerp, Emden, Bremerhaven and Hamburg show constant shares over the last decade. All of these major ports have gained the same benefit relative to their share from the extraordinary increase of the outbound transshipment. Since the emergence of a hub would have led to significant changes of its market share over time, we state that no general shift towards hub and spoke systems can be observed.

### 3.2.4 Distribution of Transshipment Volume

In a subsequent consideration, we address the share of ports with respect to the market segments short-sea, deep-sea, high and heavy, and finally second-hand car transportation for the year 2001. In summary, short-sea (feedering included) contributed with 1,810 thousand vehicles, deep-sea with 1,995 thousand vehicles, and high and heavy with 1,555

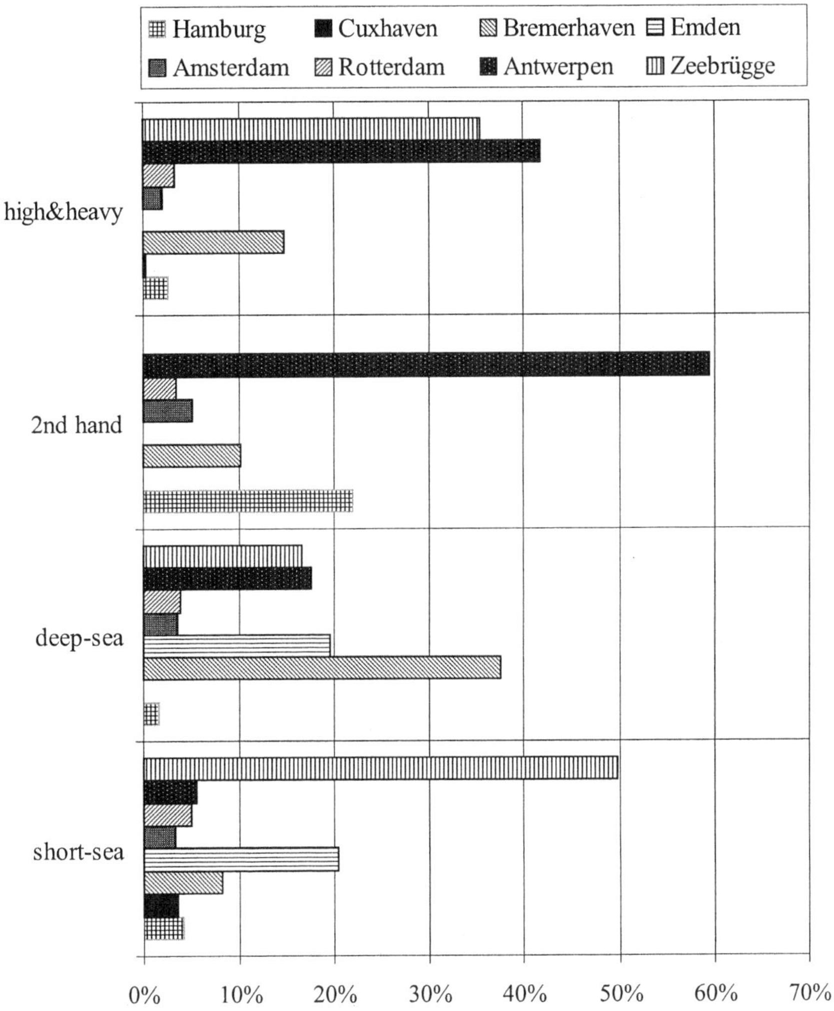


*Figure 3.7.* Percentage of transshipment volume for eight ports of the Zeebrügge—Hamburg North Range in 2001. The volume is split into the four segments "short-sea", "deep-sea", "second-hand" and "high and heavy".

thousand tons. Finally, 590 thousand second-hand cars have been exported from ports of the North Range.

Figure 3.7 shows the quantities obtained. Surprisingly, ports have adapted to high volume terminals with respect to a certain segment. This surprises, because one may have expected an evenly distributed engagement over the segments considered to minimize the risk of a shortfall concerning a single segment.

Above all, Antwerp transships almost 60% of the entire second-hand car volume of all ports. Even Hamburg holds a share of more than 20% in this segment. Additionally, these two inland ports may take advantage from local content provided by their surrounding cities.

The short-sea segment is dominated by Zeebrügge, which accounts for almost 50% ($\approx$ 1 million) of the total volume. We have already identified exports to the UK as shown in Figure 3.2 as the cause of this high share. Emden contributes with nearly 20%, mainly due to German imports from the Iberian Peninsula. Bremerhaven transships approximately 150,000 vehicles in the short-sea segment, of which the majority are feeder transports to and from Scandinavia.

Bremerhaven dominates the deep-sea segment with almost 40% of the total volume of 2 million vehicles. Emden holds a share of 20% caused by Volkswagen exports. Then, Antwerp and finally Zeebrügge follow with Far East imports of Mazda and Toyota vehicles respectively. The 'high and heavy' segment is held by Antwerp and Zeebrügge evenly balanced at around 40%. Ranking third, Bremerhaven follows with approximately 15% share. We can state from these figures that Zeebrügge, Antwerp, Bremerhaven and — to some extent — Hamburg, have successfully occupied niche markets of vehicle transshipment. Emden holds assured transshipment by means of the Volkswagen contract due to its serving a niche market.

By assuming a more or less uniform distribution of plants (vehicle sources) and dealer compounds (vehicle sinks) throughout the north-west of Europe, we must assume that from these places considerable hinterland detours (with respect to the nearest possible port connection) are accepted. This finding contradicts the interpretation of Herfort (2002a,b), who has pointed out the distance between ports and hinterland facilities as a dominant competition factor. Under this assumption a noticeable transshipment volume within every segment should be observed for every port. This contradicts our observation.

Our empirical findings are supported by Abrahamsson et al. (1998) who suggests measuring overall distances in terms of lead-times instead of kilometers or miles. Obviously, lead-times with respect to the overall distribution chain can be reduced by selecting a port that warrants a high frequency of transports. Thus, we see the primary competition factor in the frequency of transports, which entails a large transshipment volume. For short-sea ferry connections, Mangan et al. (2002) have previously identified the provision of transport capacity at a high frequency being a dominant factor in port competition.

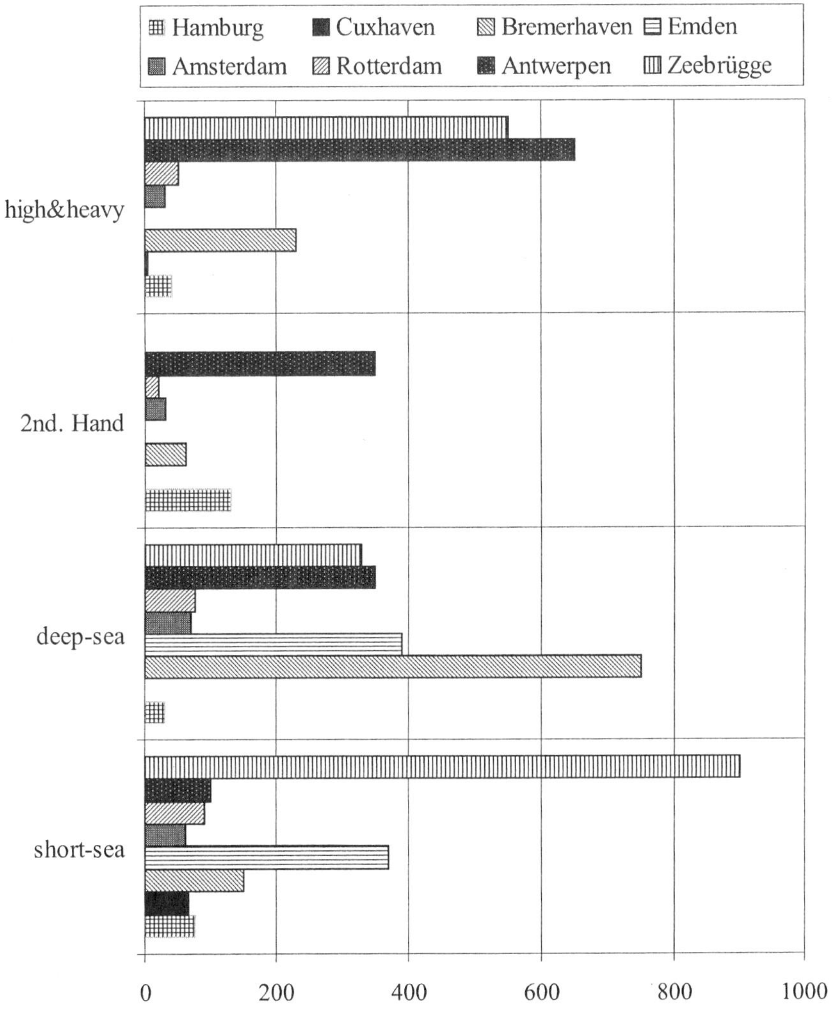


*Figure 3.8.* Transshipment volume given in thousands of vehicles for eight ports of the Zeebrügge—Hamburg North Range in 2001. The volume is split into the four segments "short-sea", "deep-sea", "second-hand" and "high and heavy".

To assess the above statements, the absolute transshipment volume in thousands of vehicles is shown in Figure 3.8 for the year 2001. Merely the numbers in the "high and heavy" segment do not refer to the transshipment volume directly, but are CEU equivalents (one high and heavy unit $\approx$ 10 CEU). The dominance of individual ports within the segments is clearly observable. With 590,000 CEU, the second hand car segment comprises a small portion of the volume of the other segments only, namely high and heavy (1,550,000 CEU), deep-sea (1,995,000 CEU) and short-sea (1,810,000 CEU). In the eight ports of the North Range a total of approximately 6 million CEU is transshipped.

Another reason for the acceptance of hinterland detouring is the demand for specialized port resources. Thus, a port's supra-structure can be the dominant competition factor for selecting a port, as is obviously the case for the transshipment of high and heavy vehicles. We see a third competition factor in the port's ability to warrant safe, reliable and efficient operations as a matter of experience, quality management and planning (Mattfeld and Kopfer, 2003). This latter factor, however, is seen as a necessary prerequisite to attract a viable transshipment volume.

Together, we expect a further increase of transshipment volume for the ports in the Zeebrügge — Hamburg North Range. Manufacturers will request logistic solutions that warrant a seamless integration of the transshipment in the individual supply chains which will enforce the demand towards specialized arrangements offered by individual ports. Since manufacturers may continue their reluctance to accept a hub and spoke system, large volumes and a high frequency of transports will remain a key advantage in the competition among ports. Once established, achieving a niche market will be unlikely to be reversed.

## 3.3 Summary

We have shown that a concentration of transshipment volume for individual ports exists, although this concentration is confined to one transshipment segment at a time. Segments are distinct markets, such that a bonus in one segment does not contribute to the success with respect to another segment. Even worse, capital bound within the infrastructure provided for one segment, may hinder investments to be taken for another segment. Finally, storage area as the probably scarcest resource cannot be assigned more than once. These are major reasons for ports to occupy niche markets.

Hinterland transportation supplies vehicles to/from the ports of high volume transshipment, and currently no extensive short-sea feedering has been ascertained. Interestingly, extensive hinterland transports are accepted in order to benefit from the high frequency of carrier callings for a port within a certain market segment. This observation indicates that further segment-oriented development of market share is possible regardless of the actual geographical position of ports with respect to vehicle sources and sinks in the hinterland.

Future developments will strengthen the role of contract logistics to implement "built to order" delivery concepts. Frequent car-carrier callings and efficient port operations are seen as a key-factor to support future "built to order" delivery of vehicles. The subsequent chapters of this book are devoted to the support of hub management by means of automated planning and scheduling. We believe that the planning of hub operations is an important prerequisite for a port in order to attend hub functionality. Thus, a strategic/tactical battle is fought on an operational battleground.

# Chapter 4

# MANAGEMENT OF TERMINAL OPERATIONS

**Abstract** The operations within a vehicle transshipment terminal can be distinguished into import and export processes. We describe both types in detail at the example of the Bremerhaven terminal, standing proxy for other large terminals with hub functionality. We then discuss methodological support for the processes on the strategic, tactical and operational level. Thereby we compare the decisions to be taken with the ones suggested for container transshipment in recent literature.

## 4.1 Transshipment Processes

Bremerhaven is one of the largest vehicle terminals in Europe (Herfort, 2002a). Its operator, BLG Logistics Group (BLG), handles approximately 1.4 million vehicles per year, compare Table 4.1 (Hafenvertretung, 2003). This tremendous volume is shipped by 850 deep-sea carriers and 500 feeder vessels visiting Bremerhaven annually (Kuhr, 2000).

For the intermediate storage of vehicles at the terminal a total of 90,000 parking slots are provided. Additional infrastructure like quays, rail ramps and truck loading areas serve as customer transfer points for the terminal. These areas are interconnected by a system of BLG

*Table 4.1.* Transshipment volumes of the automobile terminal in Bremerhaven.

| Year | **1996** | **1997** | **1998** | **1999** | **2000** | **2001** | **2002** | **2003** |
|---|---|---|---|---|---|---|---|---|
| Export | 441 | 520 | 558 | 622 | 658 | 784 | 884 | 850 |
| Import | 450 | 486 | 507 | 497 | 441 | 436 | 533 | 499 |
| Total | 892 | 1,007 | 1,065 | 1,119 | 1,099 | 1,221 | 1,417 | 1,349 |

owned travel-ways for self-propellant movements of vehicles. In order to enhance storage capacity, multi-storey park decks for more than 30,000 vehicles are provided. Next to the capacity enhancement, park decks protect the vehicles from rain and snow and ensure a high productivity of operations due to good accessibility.

Figure 4.1 presents a bird eye's view of the terminal. In the foreground we can see quays with car-carriers at berthing facilities. Separated from the waterfront by rail and ramps, there are the storage areas and multi-storey facilities. Added value facilities and transshipment points to the hinterland are located in the lower right corner as well as in the background. The land use of the terminal adds up to impressive 1.6 million square meters, such that transport distances for the transshipment processes are truly a matter of importance. In the following we briefly describe transshipment processes before we discuss management decisions on a strategic, tactical and operational level.

### 4.1.1 Import Process

Terminal operations for import vehicles cover the time-span from the ingoing call of a car-carrier to the outgoing hinterland transport. Physically we differentiate the discharge and storage of vehicles into a storage location and their retrieval from a storage location to the transfer point of outgoing hinterland transport. Both operation phases are accompanied by planning and control activities, see Figure 4.2.

Consider a Japanese car, built abroad for the German market. It is transported to a Japanese port and loaded onto a car carrier, probably in a batch with hundreds of vehicles of the same type and with the same destination. The vehicle is lashed with tie-down straps to the deck. Typically, a front to rear distance of about 30 centimeters and a side to side distance of only 10 centimeters is kept to neighboring vehicles. Thus, careful vehicle movements are essential under these narrow conditions.

While casting off to the 20 day trip towards Europe, Bremerhaven is informed about the approximate arrival at the port. About one week before berthing at Bremerhaven, the expected time of arrival (ETA) is communicated to the port authority. Additionally, the "manifesto", a freight list in the detail of vehicle identification numbers (VIN) is sent to the BLG.

Planning and scheduling of port operations can be started right after receiving the ETA. The decisions to be taken comprise the time of dispatch, the personnel engaged and the storage area to be allocated for the intermediate storage at the terminal.

After berthing, sideways ramps of the car carrier are opened, before the vehicle is unlashed by stevedores and driven to the quay side. There,

*Figure 4.1.* Bird's-eye view of the import terminal.

the vehicle is identified by its VIN and a unique bar code badge is tagged at the inner side of the front windshield. At this point of time the vehicle becomes existent to the enterprise resource planning (ERP) system of the port authority. From here, every movement of the vehicle is recorded means of portable bar code scanner and transferred to the ERP system. Finally a damage inspection is performed before the vehicle enters the terminal's storage area.

After driving a vehicle, its driver has to be brought back onto the vessel for the next vehicle to be unloaded. For this purpose the drivers are organized in groups of six, dedicated to a taxi responsible for the relocation of the group.

On average, 1,500 vehicles are dispatched from a deep-sea carrier. Typically, vehicles of the same type and with the same customer treatment customer are batched in a storage area of sufficient capacity. Since the incoming and outgoing transfer points of vehicles at the terminal are known in advance, the storage into the terminal and the later retrieval from the terminal are performed such that the driving distance covered is reasonably short.

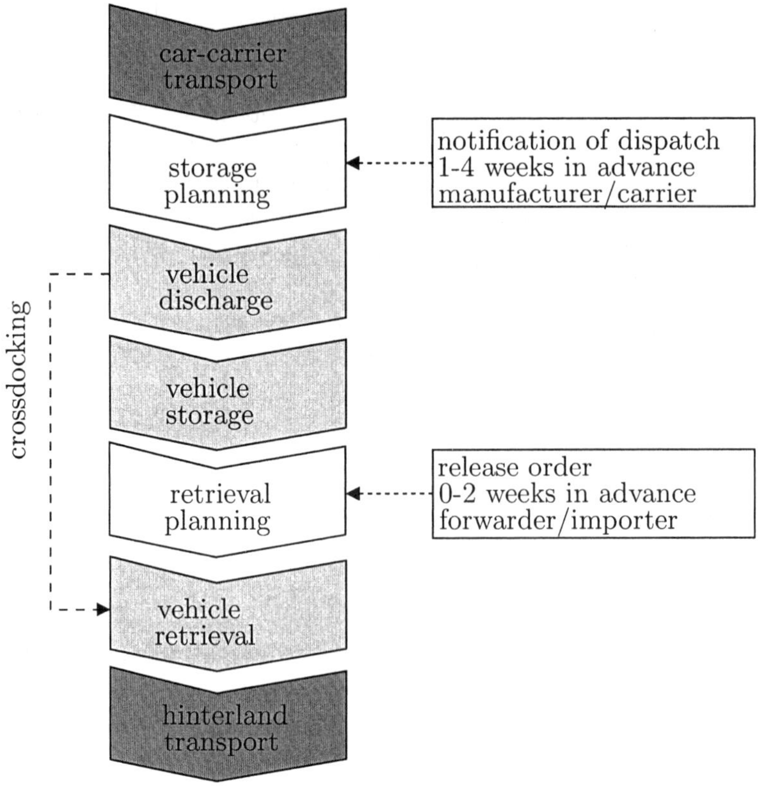


*Figure 4.2.* Business process of vehicle import transshipment.

Import partly deals with main haul runs, for which the modal shift merely entails a certain slack in the logistic chain. Since transshipment entails an intermediate period of storage, vehicles are relocated twice — denoted as storage and retrieval in the following. Typically, these vehicles leave Bremerhaven quickly, either by feeder ship or rail. According to Kiedel (2001), the minimal duration of stay at the terminal in Bremerhaven is three to four days. These cases are also referred to as cross-docking, compare Figure 4.2.

Other import vehicles are subject to complex transshipment arrangements. In particular, Far Eastern manufacturers use the terminal as a "buffer stock", because they have to supply from stock in order to compete with vehicles produced in the EU, which are increasingly "built to order". These vehicles are typically re-consolidated for retrieval with respect to a stochastic customer demand. Although the time of retrieval is not known beforehand, the transfer point provided for outgoing transport depends on the customer and the delivery mode and is therefore pre-determined.

Depending on the mode of delivery, the retrieval of vehicles from the terminal is announced up to two weeks in advance by the forwarder or import company. Thus, planning and scheduling can be done ahead of the actual retrieval operations. At the time of retrieval, the vehicles pass through a pre-delivery inspection (PDI), before they are eventually delivered by rail or truck.

### 4.1.2 Export Process

Vehicles for export arrive at Bremerhaven from inland production plants via rail or truck and remain in the terminal a few days only, before they are shipped in the majority of cases to the US. Approximately 85% of the outgoing vehicles come by rail in block trains of a minimum length of 600 meters consisting of at least 17 freight cars. German rail offers a priority service for this type of transport such that a maximal transport time of 24 hours within Germany is guaranteed. The export quantities and the associated production dates of vehicles are subject to long-termed production planning. Since carrier capacity is allocated in advance of the vehicle production, vehicle transshipment is a matter of sorting and merging incoming vehicles into storage areas dedicated to already scheduled car carrier departures.

A dedicated area is usually formed by one or more parking lines each of them sufficient to hold dozens of vehicles. Vehicles arriving in a block train are provided for different destinations. These destinations can be served either by distinct liner services or by the same carrier in a prescribed order of port calls. In the latter case, the stowage plan for the car carrier has to respect the order of vehicle dispatches in the sequence of port calls. Vehicles arriving at Bremerhaven for export have to be assigned to parking lines such that the stowage plan of the outgoing vessel is supported by the parking line assignment.

At the time the car carrier is arriving in Bremerhaven, the loading operations have to be performed efficiently. Kiedel (2001) reports a case of the carrier 'Kassel' where the unloading/loading operation of 5.300 vehicles has been performed in 5 working shifts. This figure yields an impressive loading time of 5 seconds per vehicle operated. The provision of efficient loading operations is the primary challenge of operations planning for export transshipment.

### 4.1.3 Auxiliary Process

Due to the stochastic retrieval process of vehicles stored in buffer stocks, the utilization of storage areas tend to get fragmented over time. Relocations of vehicles may be required in order to warrant an efficient en bloc reuse of storage areas. However, additional relocations have to be carefully planned and kept at an absolute minimum.

Because of the risk of damages to vehicles whilst being moved manufacturers are unwilling to accept moves other than necessary for storage and retrieval. Furthermore, new vehicles should have driven a distance of less than 10 kilometers when handed over to the customer. Thus, self propellant movements at the terminal are to be confined to the minimal distance only.

## 4.2 Management Decisions

The design of terminal processes is a matter of management decisions. As yet, there has been almost no methodological support available for vehicle terminal operations of the type already developed for container transshipment (Vis and de Koster, 2003; Steenken et al., 2004; Günther and Kim, 2004). So far, only Mattfeld and Kopfer (2003) propose an operational model for the planning of vehicle transshipment tasks and Fischer and Gehring (2004); Fischer (2003) proceed this research into vague direction when compared to the problem as it appears in practice.

With regard to the scope of decisions to be taken we can differentiate strategic from tactical and operational issues. Usually, decisions on the strategic, tactical and operational level are interrelated. For example, the strategic decision to seal an area for storage, the tactical decision for a certain layout of the storage area and operational decisions of utilizing the storage area all contribute to a common measure of terminal operations.

A typical measure for the quality of transshipment processes is the productivity of operations, i.e. the mathematical inverse of the production coefficient as known from operations management. For vehicle relocation processes the productivity measure denotes the number of vehicles moved within a certain time span by a driver. For instance, an import process run at a productivity of 4.5 means that on average a single driver moves 4.5 vehicles per hour into a storage area.

The measure expresses customer oriented service capability and cost oriented efficiency at the same time. The measure can be easily recorded in practice and it allows a meaningful interpretation as aggregate measure as well as for detailed operations. Surely, a tradeoff between a high productivity and a low process quality e.g. in terms of vehicle damages exists. Since customers nowadays expect an almost zero damage rate, management decisions address the maximization of productivity under the constraint of an almost perfect customer service.

### 4.2.1 Strategic Decisions

Strategic decisions focus on business strategies on a time horizon of years. Such long-term oriented strategies effect the infrastructure and the processes within the whole port. Strategic decisions are hardly irrevocable and are seen as a prerequisite for future business success.

The probably most important issue is the choice of a strategy to develop one segment of the vehicle transshipment market as a strategic business unit. Because of scarce resources, a strategy will be taken up in favor of further opportunities in other market segments. I.e. focusing on the high & heavy segment requires infrastructure investments in sealed ground and armed loading ramps in order to attract a high frequency of carrier callings at the port. Typically, vehicle transshipment segments compete with one another as well as with other intermodal transshipment markets like container or common cargo transshipment. Thus, a strategy supporting the high & heavy segment will make the port less attractive for other transshipment segments like deep-sea, short-sea or used cars.

As we have shown in Chapter 3, a strategy aiming at a high market share within a dedicated market segment has already been adopted by all major ports of the North Range. This is seen as the main competitive advantage for safeguarding business (Porter, 1980). A strategic business unit is implemented by investments with regard to the infrastructure and to the business competence. Infrastructure decisions comprise the provision of sealed ground, travel-ways, berthing facilities and park decks. Also excellent rail and highway connections to the hinterland have to be provided. For ports located close to the coastline a sluice system will help to keep loading and unloading operations independent of the tide. This can enormously speed up ship service times and is therefore regarded as a key-feature in regions of high tide.

Business competence in terms of process orientation, quality management and human resource development becomes increasingly important also for the traditionally oriented logistics industry. Port operators start to act as third party logistics provider for vehicle manufacturers and offer added value and vehicle inspection services besides traditional transshipment. The intention behind such activities is to bind customers with integrated logistics functions into long lasting contracts. Such functions also include integrated information systems with a tracking and tracing capability along the entire supply chain.

Although in theory strategic decision-making can draw on methodological support offered by standard approaches to hub location (Domschke and Krispin, 1997; Racunica and Wynter, 2000; Goetschalckx et al., 2002), in practice vehicle transshipment terminals are typically built in an already existing port infrastructure. Even if in rare cases terminals are newly built like recently in Zeebrügge, the number of potential locations is very small and only a few alternatives have to be considered. Different to distribution terminals located in the hinterland, for ports location models play a minor role only.

To summarize, strategic decisions aim at implementing a strategy to attract business to the port (Paixão and Marlow, 2003). Investments in infrastructure and business competence are necessary prerequisites for successful negotiations with carriers and manufacturers. Once a port is integrated in a carrier's liner service, contracts are awarded on a time horizon of years. It is the challenge of tactical management decisions to provide efficient operations in this given strategic framework.

### 4.2.2 Tactical Decisons

Tactical Decisions address agreements, roles and procedures in order to translate strategies into operational issues. To this end tactical decisions interface long-termed strategies and short-term planning and scheduling with decisions on a mid-termed horizon. Thus, the goal at the tactical level is to manage premises for the control of resources on a day to day level.

**Ship Scheduling.** Carrier lines operate in accordance with a fixed schedule. Given the expected vehicle volumes to be loaded and unloaded at the ports under consideration over a mid- to long-termed horizon, the liner schedule can be derived from a capacitated pickup- and delivery problem (Ronen, 1993; Fagerholt and Christiansen, 1999; Bendall and Stent, 2001; Mourão et al., 2001). Vehicles to be loaded at ports correspond to pickup operations whereas vehicles to be dispatched are depicted by delivery operations. The distance in miles between ports is divided by an average carrier speed of 20 knots leading to approximated shipping times between the ports involved. By assuming standard productivity rates for loading and unloading operations, port service times can be determined. The ship scheduling problem aims at determining a ship route and a service schedule which maximizes capacity utilization. However, the ship scheduling takes place at the carrier line and is typically beyond the scope of the transshipment terminals.

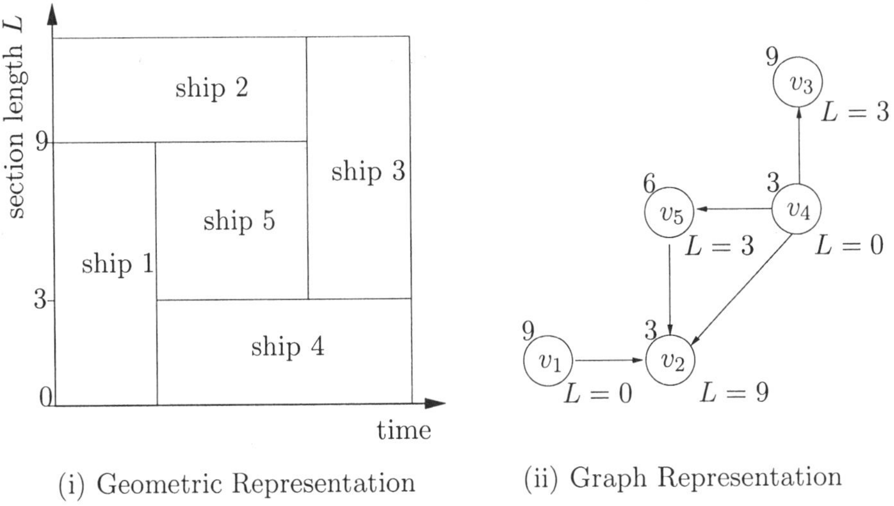


*Figure 4.3.* Representations of berth planning problem.

**Berth Allocation Planning.** A schedule prescribes the arrival time and approximated service time at a port. All calling ships are to be serviced without waiting times. Tidal currents and restricted sluice times introduce a structure of "berthing time slots" in the granularity of hours easing the temporal assignment of car carriers to berthing facilities. However, multiple berthing locations exist (compare Figure 4.1) and ships have to be assigned to distinct sections of a berth. In the context of container transshipment Nishimura et al. (2001) consider the distance between the berth and the relevant container yard by means of a variable handling time $C_{ij}$ of ship $i$ at berth $j$. The authors minimize the sum of waiting and handling times over all ships to be serviced. In vehicle transshipment the transfer points to the hinterland transport are known in advance, so we can proceed in a similar way. Berth facilities are chosen such that the expected routing of vehicles through the terminal becomes minimal (Hall, 1987).

Car carriers are of different length. A berth planning approach proposed in Lim (1998) takes this fact into account. The author proposes a model and suggests a heuristic procedure to solve the problem. In the following we sketch (a simplified version of) the model in order to outline the characteristic of the berth planning model. A geometrical representation of the problem depicts a solution in a rectangular plane. The x-axis represents the time whereas the y-axis depicts the length of the berthing facility. A rectangular plane describes that a ship occupies a certain section of the berthing facility for a certain time span.

In Figure 4.3, first ship 1 and ship 3 share the berthing facility. After ship 1 has left, ships 4 and 5 take over the free section of the berthing facility. Ships 2 and 5 leave the berth simultaneously later on, so that ship 3 calls at the free part of the berth. Lim suggests a graph representation where nodes $v_1 - v_5$ represent the berthing of ships. The section length required for berthing is given as node attribute. Lim starts by adding undirected edges between nodes representing ships with overlapping berthing time, i.e. $(v_1, v_2), (v_2, v_4), (v_2, v_5), (v_3, v_4)$, and $(v_3, v_4)$. Then, the undirected arcs are successively turned into directed arcs representing the order of berthing at the mooring facility. The heuristic aims at an acyclic graph which minimizes the length of the longest path. Obviously, the longest path in the graph corresponds to the maximum usage of berthing facility $L$ over time.

For the berth planning problem in vehicle transshipment we suggest to apply the decision version of the above optimization problem, i.e. to find a directed acyclic graph where no path exceeds a given length $L$ of the berthing facility. In order to keep the handling times reasonably short (corresponding to a high productivity), we suggest to search the graph satisfying the above constraint while minimizing the sum of expected handling times as suggested by Nishimura et al. (2001). However, the handling times can be taken as a rough estimate only, because the actual number of vehicles and the particular routing through the terminal are not known in advance. Anyway, a mid-termed berth allocation planning on the basis of scheduled car-carrier lines will support terminal operations to a large extend.

**Storage Space Partitioning.** Within a given infrastructure determined by strategic planning, berthing facilities, storage areas, travelways and transfer points can be devoted to particular tasks on the level of mid-termed decisions. As outlined in Section 4.1, the types of vehicle transshipment can be distinguished into batch processing and individual transshipment arrangements. Dedicated routes through the terminal can be determined between berthing facilities and transfer points to the hinterland. If for instance a Japanese manufacturer transships large batches of vehicles via rail to East Europe, the terminal layout should allow a direct routing with a minimal distance through the terminal. The provision of suitably partitioned storage space should exist on the straight line connecting the rail ramp and its closest berthing facilities.

However, in a large terminal many such interleaving routings exist. Thus, the assignment of dedicated storage areas to distinct vehicle routes may lead to an inferior overall performance only (Goetshcalckx and Ratliff, 1990). The authors favor shared- over dedicated storage assignment policies for warehousing. In the context of automated storage/retrieval systems, Muralidharan et al. (1995) propose a shuffling heuristic combining class based (i.e. shared) and random assignments to storage locations. With this approach service times can be significantly reduced. Petersen (1999) stress the interdependency of routing- and storage policies for warehouse efficiency. They argue that for a given storage policy routes produced by simple heuristics are much more intuitive for implementation than routes produced by optimizing algorithms.

Simulation models help to determine a suitable structuring of the storage policy with respect to the actual needs. See Yun and Choi (1999); Chen (1999); Shabayek and Yeung (2002); Murty et al. (2005) for references on simulation approaches for container terminals. For vehicle transshipment terminals the partitioning of the overall storage space in areas of suitable size and location is of predominant importance. Typically, this mid-termed decision has to be taken in accordance with existing transfer points and tavel-ways, compare Figure 4.1. For the purpose of operations planning in the Bremerhaven terminal, a partitioning of the storage space has led to 80 areas of approximately 1000 vehicles capacity each, see Figure 8.2. In a next step, for each of the storage areas a suitable storage layout has to be determined.

**Storage Area Layout.** The layout of storage areas interfaces infrastructural decisions with the options to be taken by operational planning of vehicle transshipment. As we have already seen, import and export processes differ in the way vehicles are stored into and retrieved from a storage area. Regarding import, vehicles are stored as a batch and contingent of the distribution policy, vehicles are either retrieved in the same batch or they are subject to re-consolidation due to stochastically arriving customer delivery orders. For export, vehicles are consolidated according to their destination in order to warrant an effective car carrier loading in a batch.

The layout of the terminal's storage space has to support all of the above processes, which will hardly be obtainable by just one storage layout. To cover the distinct needs, the layout of storage areas provided for processing batches is distinguished from a layout suitable for stochastic retrieval. Batch processing can be best supported by dividing a rectangular area by lines, such that vehicles are queued within the resulting stalls in a first-in-first-out (FIFO) fashion. The length of the stalls should be chosen in accordance with the mean size of batches occurring such that the overall space utilization is sufficiently high. Two contradicting issues determine the width of stalls. A small width contributes to a higher utilization of the storage space but will come along

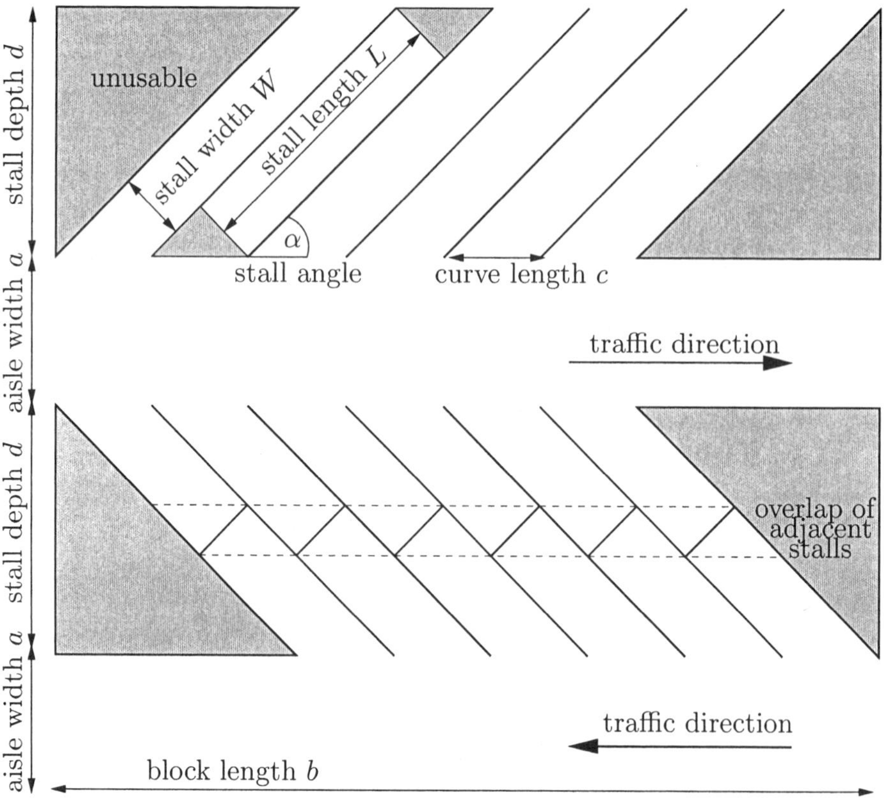


*Figure 4.4.* Aisle and two rows of stalls at angle $\alpha$.

with an increased damage rate of vehicles. Therefore quality considerations of vehicle manufacturers will typically dictate minimum measures for the dimensions of stalls.

Although large volumes of vehicles are processed in batches, the space requirement for this type of transshipment is rather small. The mean dwell time of vehicles in the terminal for this type of transshipment will be in the range of a few days only. To the contrary, import vehicles in buffer stocks remain longer in the terminal and consequently a major portion of storage space is devoted to these processes. Therefore, the storage layout for buffer stocks is of predominant importance. In the event that vehicles are stored in lines, only the first and the last vehicle can be accessed directly, which complicates the re-consolidation of vehicles for retrieval. With increasing line length $n$, the number of directly accessible vehicles decreases with $2/n$, and on average $(n-2)/4$ vehicles have to be shunted in order to access a stored vehicle (provided that the

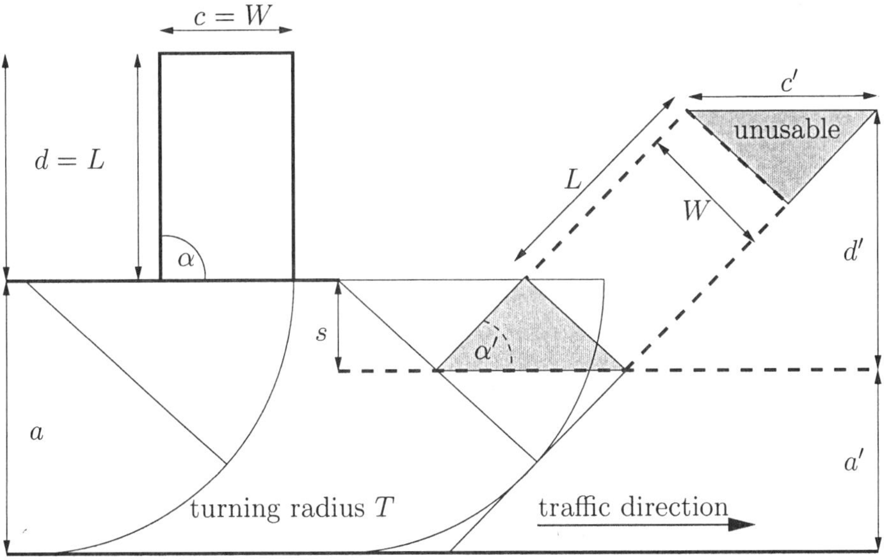


*Figure 4.5.* Saving of aisle width due to a stall angle $< 90°$.

capacity of the storage area is exhausted). In order to avoid shunting at all, $n = 2$ is required.

The stochastic retrieval process for buffer stocks suggest stall length of $n = 2$ which increases the space required for aisles in order to access the stalls. The topic of maximizing space utilization for a parking lot has been discussed in literature, compare Cassady and Kobza (1998); Thompson and Richardson (1998). For quantitative approaches in parking lot design see Bingle et al. (1987); Iranpour and Tung (1989). In the following we survey the approaches proposed in the latter two references. We consider a flat and rectangular storage area where stalls for head-in parking are to be arranged in a symmetric way. We assume that the stalls are accessible through parallel one-way traffic aisles, see Figure 4.4. Given a prescribed number of storing positions, we aim at minimizing the storage space required.

We consider the stall width $W$ and the stall length $L$ to be constant, and decide upon the angle $\alpha$ including the orientation of the aisle and the stall. The stall depth $d$, the curve length $c$ and the block length $b$ depend on $\alpha$. For $\alpha = 90°$ the storage space is completely utilized, which appears to render a minimal overall space requirement. The smaller $\alpha$ gets, the larger the unusable space (grey shaded in Figure 4.4) gets. However, the overlapping of stalls as shown in the lower part of the figure do not contribute to the unused space.

Unfortunately, for $\alpha = 90°$ one quarter of the circular arc with turning radius $T$ is needed in order to access the stall, compare the left part of Figure 4.5. For $\alpha' < 90°$ the turning angle required while entering the stall becomes smaller, because from a certain point on the vehicle can continue on a straight line. This point is determined where the extended outer diagonal of the stall forms a tangent with the turning arc. Because of the smaller turn, the width of the aisle $a'$ can be diminished by $s$. Bingle et al. (1987) show, that there is a tradeoff between increasing $c, d$ to $c', d'$ and decreasing $a$ to $a'$. For large parking lots with the same number of places, the calculations yield a smaller overall area consumption for $\alpha = 74.2°$ than for $\alpha = 90°$. Iranpour and Tung (1989) confirm this result and prove, that an optimal $\alpha$ with respect to $W, L$ and $T$ exists.

Next to a saving of storage space, $\alpha < 90°$ alleviates the likelihood of damages due to turning operations. This complies with the quality requirements of the vehicle manufacturers. Unfortunately, a smaller angle is not applicable with larger stall length. These have to be operated in a FIFO sequence, i.e. head-in parking and back out leaving is no longer feasible. Instead, leaving a stall in the front direction of the vehicle requires a turning arc $> 90°$ in order to comply with the driving direction of the opposite aisle, compare the bottom part of Figure 4.4. Thus, for batch operations we commend large stalls operated in a FIFO way with stall angles of 90°. For import processes, which require an individual access to vehicles for retrieval, we commend small stall length and stall angles of approximately 70°.

**Personnel Planning.** The transshipment of vehicles is a manpower intensive task. Different to container transshipment, operations cannot be automated by means of automated guided vehicles, see Section 3.3. in Steenken et al. (2004). On a strategic level of planning, a structure for the recruitment and training of personnel has to be set up. On a tactical level, a working shift model has to be determined, which allows the flexible usage of personnel depending on the scheduled carrier calls at the port.

At the Bremerhaven terminal, in former years workload peaks have been resolved by a flexible hiring of personnel from external sources, i.e. by workers from a port-wide workforce pool or by volunteers from the local auxiliary fire brigade. Advances in quality management now restrict the recruitment of driving personnel to well trained employees with regular work contracts. In order to keep the flexibility of workforce, suitable labor tour models can be applied (Brusco and Johns, 1996; Aykin, 2000). Given a dynamic workforce demand derived from a

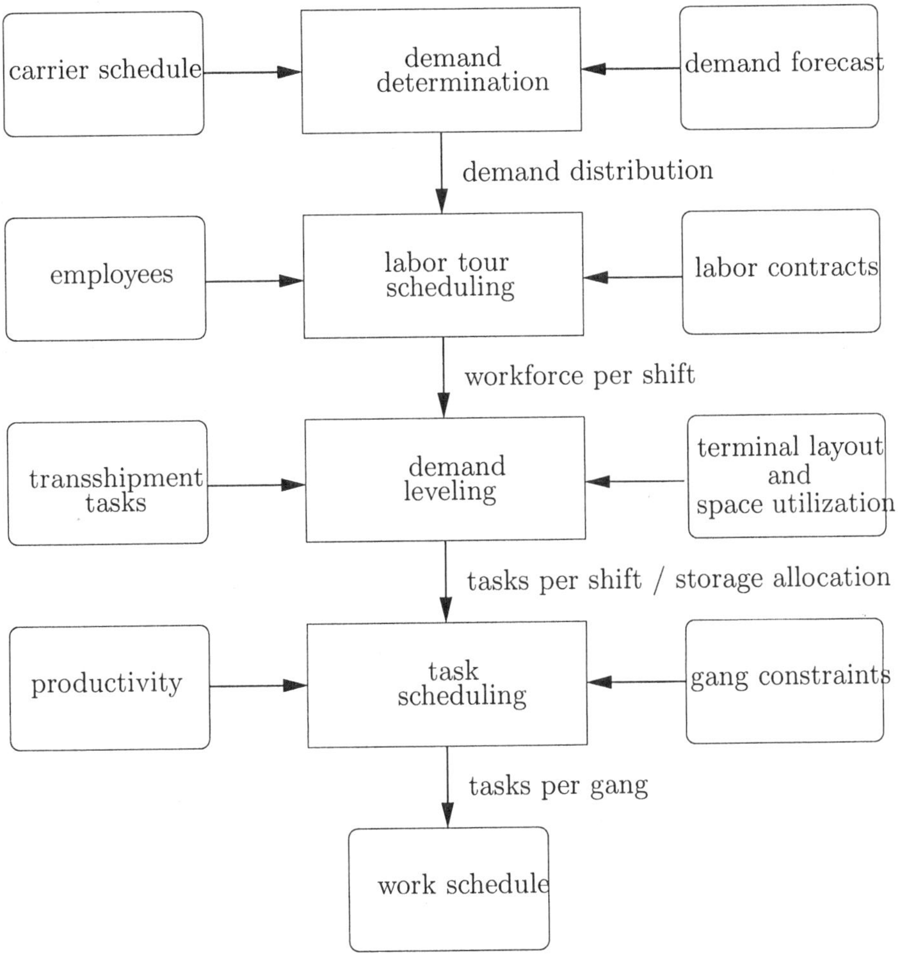


*Figure 4.6.* Mid- and short-termed planning and scheduling problems.

mid-termed liner schedule, labor tour scheduling arranges work shifts of personnel in accordance to the expected demand distribution.

Nobert and Roy (1998) illustrate the interdependency between labor tour scheduling and demand leveling in order to cope with an unevenly distributed workforce demand. The authors consider the planning and scheduling of transshipment personnel at an airline cargo terminal. Despite the fact, that many handling tasks have to be executed on time, other handling work can be deferred. Thus, a balancing of workforce demand is performed either by providing extra personnel at demand peaks, or by shifting the demand to non-peak times. Due to the stochastic environment, Nobert and Roy (1998) perform both planning tasks simultaneously at the stage of operational planning.

For vehicle transshipment, the fixed liner schedule and the mid-termed announcement of vehicle transshipment allow to perform the labor tour scheduling on the level of mid-termed planning, see Figure4.2. Demand leveling and detailed scheduling are performed later on at the level of operational planning. Figure 4.6 surveys the steps of integrated personnel and transshipment task planning. In advance of operations, the liner schedule and demand forecasts are used for the determination of a future demand. Labor tour scheduling takes the demand distribution, the number and attributes of employees to schedule and further constraints due to labor and union contracts. Labor tour scheduling aims at an assignment of personnel to work shifts such that the expected demand distribution is met as close as possible.

The planning and scheduling of actual transshipment operations is deferred to decision-making on the operational level.

### 4.2.3 Operational Decisions

At the operational level, decisions focus the seamless and efficient implementation of work processes on a short-term horizon. In contrast to the process oriented view taken in Section 4.1, the integrated view to resources is of particular importance. The allocation of storage space, the scheduling of transshipment operations and the assignment of personnel to operations are important decisions to be taken.

**Storage Space Allocation.** For the related problem of container transshipment, approaches aim at improving the performance of operations by means of transport control and storage space allocation. A high productivity of the gantry cranes shortens the berthing time of vessels, which is the most important goal for container transshipment. Transport scheduling and storage space allocation aim at avoiding bottlenecks in the service of gantry crane operations. Bottlenecks can be avoided by using marshaling yards where containers are sorted prior to gantry crane operations. Another strategy aims at a spread of containers to (and from) multiple storage locations in order to alleviate congestion of transport vehicles at the storage yard.

For both strategies, a tradeoff between the transport time and the handling time has to be taken into account. Central locations accessible at a high productivity will be heavily used which leads to the stacking of containers. Stacking in turn lengthens the access time for containers which is referred to as handling work. Thus, it may be advantageous to favor remote locations for storage and to avoid extra handling work this way. Since information about the hinterland of transport is typically lacking, see Steenken et al. (2004), storage allocation merely supports gantry crane operations rather than providing a distance minimal routing of containers through the terminal.

Approaches to the combined transport and storage allocation problem are reported in Sagan and Bishir (1991); Taleb-Ibrahimi and Castilho (1993); Steenken et al. (1993); Kim and Kim (1999); Holguìn-Veras and Jara-Dìaz (1999); Kozan and Preston (1999); Böse et al. (2000); Bish et al. (2001); Preston and Kozan (2001); Zhang et al. (2003); Kim and Park (2003); Ebben et al. (2004); Nishimura et al. (2005).

Terminal operations in vehicle transshipment differ significantly from operations in container transshipment.

- First, container flows are strongly fragmented and planning is done for the entity of a single container. Vehicle flows have much in common with bulk cargos. The notion of bulk grouping allows the definition of reasonably sized entities for planning.

- Second, containers may be relocated several times during their stay in a terminal. Due to the danger of damage resulting to vehicles, the practice of relocation is avoided at vehicle terminals. Since the relocation of vehicles should be kept to a minimum, their assignment to appropriate locations is a matter of importance.

- Third, containers can be stacked upon one another, increasing storage space, whereas vehicles cannot. The area taken up by vehicle stocks is enormous, so that the distances to be covered become an important factor in the planning process.

- Finally, different to container transshipment, reliable data about operations is available in advance. In particular, the locations vehicles are entering and leaving the terminal are known.

We see the above properties as prerequisites for an effective methodological support of transshipment operations. Particularly the grouping of vehicles into transshipment task entities allows to base operational decision on optimization. In actual fact, the detailed usage of resources can be planned days ahead of operations by means of a mathematical optimization model.

Tactical decisions provide berthing positions or carriers, storage areas with a prescribed layout, and a personnel capacity n the granularity of work shifts. Goal of decisions at the operational level is the balancing of work load on a detailed level, such that the personnel capacity is met and operations are performed at a high productivity. Planning is performed on a rolling horizon allowing that capacity bottlenecks can be identified and alleviated an an early stage.

To adopt the workload to the prescribed level of personnel, two opportunities can be taken. First, one portion of transshipment tasks can be shifted back and forth in time. Second, storage locations for vehicles can be chosen such that movements to nearby locations at a high productivity are performed in congested work shifts whereas remote locations are favored under a relaxed workload situation. Generally, a placement of vehicles at an area is chosen, which is located close to the straight line connecting the given two transfer points of a transshipment task. In case of congestion, a nearby storage location is chosen such that the distance to be driven is reasonably small, i.e. the task is processed at a high productivity.

Since storage space is a scarce resource, a capacitate multi-period optimization model is required in order to benefit from operational decisions. Its in- and output is sketched in the lower part of Figure 4.6: The personnel assigned to individual work shifts is received as input from the mid-termed planning. Additionally, the capacity utilization of the storage system and the distance between areas of the system are input to the demand leveling step. Finally, transshipment tasks are defined by grouping vehicles with the same manufacturer and the same destination.

The demand leveling assigns tasks to work shifts and storage space to tasks, such that the prescribed level of personnel is met and transshipment tasks are performed with reasonable high productivity. Chapter 5 presents the optimization model and Chapter 6 presents an Evolutionary Algorithm to solve the complex problem at hand. So far, a detailed assignment of transshipment tasks to (groups of) drivers is missing. This is subject to the final scheduling step.

**Gang Scheduling.** A transshipment task is processed by a dedicated group of driving personnel. Dependent on the charge of vehicles and the distance moved, a number of taxis carrying up to six drivers each form a gang with a dedicated foreman. Since for example carrier unloading is performed under narrow conditions in the vessel, for this type of task the maximum size of a gang is limited to approximately 50 drivers. At the minimum, only one taxi with even less than 6 drivers can form a gang.

Because of organizational issues, a gang structure is fixed for an entire work-shift of 7.5 hours. Neither the number of gangs nor the number of drivers assigned to a gang is given in advance. Thus, first a suitable gang structure has to be determined. Then, the transshipment tasks are to be assigned to the individual gangs.

Earliest starting times and latest finishing times complicate the seamless integration of tasks into a gang structure. Additionally, precedence constraints between transshipment tasks apply. The optimization model on this detailed level of operations aims at the minimization of driving personnel in order to perform a set of transshipment tasks within a work shift. It takes the level of personnel from the demand leveling step as an lower bound and proceeds by keeping the number of additional drivers as small as possible. A deviation from the lower bound means that waiting times for gangs occur because of logical or temporal constraints.

The in- and output for this detailed scheduling step is sketched in the bottom part of Figure 4.6. The storage allocation already performed in the demand leveling step is replaced by a productivity measure corresponding to the distance covered. In this way the storage area utilization and the distance between these areas is excluded from the scheduling model. Gang constraints are taken as a further input such that eventually a detailed work schedule is determined. Again, Chapter 5 presents the model and Chapter 7 deals with an elaborated neighborhood search technique developed for the problem.

**Operational Issues.** Operational issues refer to details performing transshipment tasks. This includes, for example the loading and unloading of auto trucks, rail wagons and car carriers (Agbegha et al., 1998). Furthermore, the placement of vehicles inside a storage area is of concern. So far, storage space has been allocated merely on the level of storage areas (of approximately 1000 parking stalls each). In even more detail, the utilization of an area with respect to its storage layout has to be determined. This is done by priority rule based assignment during the execution of operations. Although certainly experience is needed to implement work processes with a high productivity, these issues are not further investigated throughout this book.

## 4.3 Summary

In this chapter transshipment operations within a terminal have been described twice. First, a process oriented view has been taken in order to sketch the work-flows of interest. Then, a resource oriented view has been taken in order to depict the interrelations of resources shared by the work processes involved. The latter view has been divided with respect to management decisions on a strategic, a tactical, and an operational level. The differentiated view allows the optimized assignment of resources by means of methodological support. In the next chapter we focus on the operational level of planning and scheduling in detail and propose a mathematical optimization model. This is operationalized

into the demand leveling and scheduling step as already done informally in this chapter.

Chapter 5

# MODELING TERMINAL OPERATIONS

**Abstract** In this chapter we describe terminal operations to be performed in vehicle transshipment. Particular attention is paid to the key-factors for efficient operations. We show, that next to the matter of a high productivity, the balancing of the operation effort is an important goal of terminal operations. A balanced load ensures safe and reliable operations, a prerequisite for the transshipment of finished vehicles. Finally, we discuss tasks and objectives of automated planning and scheduling for vehicle transshipment.

We formalize operational issues of vehicle transshipment in a mathematical model, which integrates storage space allocation and personal deployment. Such a model is needed in order to assess the problem difficulty with respect to the number and type of variables, objective function and constraints. Furthermore, the model supports the organizational embedding of decisions along the planning process. Since the problem is dynamic by definition, time and/or conditions have to be determined at which recent planning is to be renewed. Finally, the model is useful to derive manageable problems from a hierarchical problem separation. To this end the integral model is separated into two subproblems coupled by an anticipation term.

## 5.1 Vehicle Transshipment Planning

In this chapter we present a mathematical model for the general case of integrated space allocation and personnel deployment. Although developed for vehicle transshipment, the model is applicable to a wide array of transshipment applications. The model covers the transshipment of non-homogeneous goods, which require an intermediate storage at a terminal. Reasons for storage comprise a modal shift of goods, the need for a certain slack in the supply chain, or the need for a buffer stock in order to ensure delivery reliability.

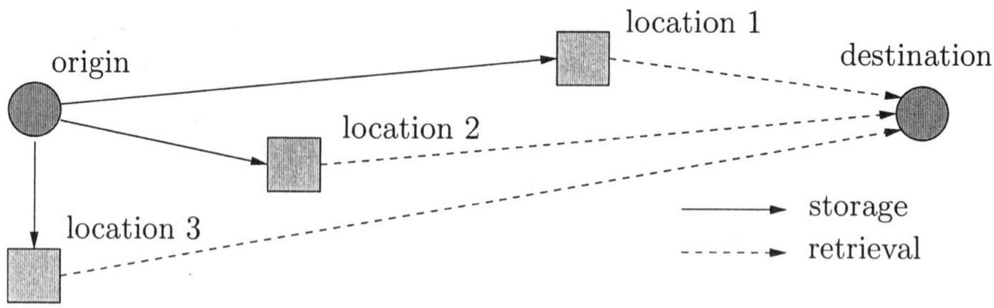


*Figure 5.1.* Three alternatives of allocating a storage location are depicted.

### 5.1.1 Allocation of Storage Space

Central to our approach is the notion of a *task*. A task comprises the relocation of a number of identical (assumed) vehicles, which are treated as bulk cargo. The vehicles included in the task are supposed to be transported from an origin to a destination in a given, typically narrow time window. We differentiate between "storage tasks" entering vehicles to the terminal and "retrieval tasks", performing the vehicle dispatch from the terminal.

We refrain from considering actual distances between locations. Instead we consider a productivity measure, i.e. the number of vehicles that can be moved between two locations per unit time. This measure is based on distances, but includes setup times and may even be modified in order to incorporate bottlenecks in the travel way system, etc.

A pure modal shift consists of two successive tasks comprising the same volume of vehicles. If intermediate storage beyond the planning horizon is required, storage and retrieval tasks are handled independently. The same treatment applies for vehicles to be kept in buffer stocks. Here, a single storage or retrieval task depicts the consolidation into a storage area, or the vehicle commission from a storage area.

Customers do not necessarily insist on the transportation within a certain period, but grant time-windows for both relocation types, the storage into the transshipment hub and its corresponding retrieval. In order to model that the storage and retrieval may fall asunder, we consider the transshipment of a charge of vehicles as a pair of storage and retrieval tasks coupled by a precedence constraint.

Whenever customer granted time-windows for a pair of tasks do not overlap, an intermediate storage becomes unavoidable. In these cases a storage location of sufficient capacity is chosen, such that the manpower demand for both logistic tasks, storage and forthcoming retrieval, is reasonably small. Whenever the time-windows for a pair of storage and

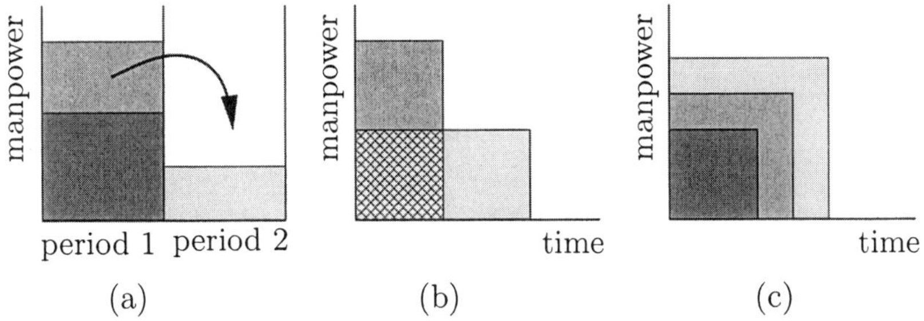


*Figure 5.2.* Illustration of options to balance manpower. Tasks can be shifted between periods (a), performed in different modes due to a trade-off between time and personnel requirement (b), or due to a change of productivity due to the choice of a storage location (c).

retrieval tasks overlap, typically both tasks are assigned to the same period, i.e. the vehicles are shipped directly. The consumption of storage space can be neglected in this case.

A small overall inventory level allows the greatest choice among storage locations; therefore typically storage tasks are assigned to their latest permissible period of processing, whereas retrieval tasks are correspondingly assigned to their earliest period. Besides a small overall manpower demand, balanced manpower utilization over the periods considered is an important goal of the terminal management. A balanced load ensures safe operations and simplifies the integration of prioritized tasks that arrive late.

Therefore an earlier period of storage and/or a later period of retrieval can be advantageous. Even if direct shipment can be performed, an intermediate storage of vehicles may be preferable in order to balance manpower.

Although differently skilled personnel work together in order to perform a task, we focus on the drivers, whose costs are almost proportional to the number of vehicles moved and the distance covered. Therefore, storage locations are assigned to tasks such that the overall distance of storage and retrieval is minimized. Even if equal overall distances are considered, the distribution of storage locations has a significant impact on the manpower usage. Under a congested condition, we prefer storage into a nearby location. In this way the utilization of driving personnel is kept low at the expense of a higher driver demand for the future retrieval.

Figure 5.1 illustrates the dependencies. One may accept reasonable higher manpower consumption — resulting from a long distance — for periods of modest manpower utilization. This decision turns out advantageous if it yields a short retrieval distance for a forthcoming congested period, compare 'location 1' and 'location 2'. In extremely congested periods, 'location 3' may even be preferred — despite the longer overall distance — because of its exceptionally short storage distance.

### 5.1.2 Manpower Deployment

In order to ensure safe and reliable operations, drivers are grouped into gangs of between 5 and 50 drivers assigned to a dedicated foreman. In this way the choice of modes of performance can pursue the seamless integration of tasks into a gang structure. However, neither the number of gangs per shift nor their sizes are known in advance. In actual fact, gangs are set up flexibly depending on the characteristics of the tasks to be performed during a shift. Thus, scheduling pursues both, determining a gang structure and fitting the tasks into this structure.

Figure 5.2 illustrates options to perform tasks. First of all, tasks can be moved between shifts if permitted by their time-window of processing (a). Furthermore tasks can be performed in different modes resulting from a trade-off between driving personnel and processing time required (b). The interdependencies of location assignment and personnel usage are shown in (c): Since the (location dependent) productivity has to be substituted by driving personnel and vice versa, the location capacity planning and the detailed task scheduling are linked into one integral problem.

The performance of tasks is constrained in multiple ways. First, a storage task must be performed before its corresponding retrieval can take place. In addition bottlenecks with respect to storage capacity have been taken into account between otherwise unrelated tasks. For example, a retrieval task has to be completed clearing a storage area, before another storage task can be processed.

Luckily, no upper limit for the usage of manpower has to be taken into account because drivers can be hired flexibly from a port-wide workforce pool. Management aims at avoiding short-term hiring due to the fact that inexperienced drivers tend to increase damage rates and decrease productivity. Therefore, besides efficiency issues, an evenly balanced allocation of manpower is pursued.

### 5.1.3 Goals for Terminal Operations

In the first case the safe and reliable accomplishment of terminal operations has to be ensured. Although carrier, forwarder and hub operators are primarily interested in a high productivity of operations, all of them have to meet the concerns of the manufacturer and avoid damaging of

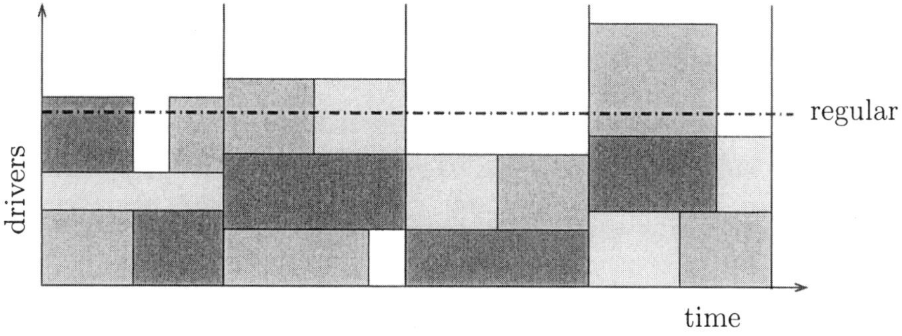


*Figure 5.3.* Illustration of the objective function considered.

vehicles whenever possible. However, an unbalanced workload will result in congested periods with a relatively high risk of damages. Thus, terminal operations aim at balanced manpower utilization under the constraints of efficient loading and unloading operations.

Obviously, it is trivial to balance workload over the shifts by decreasing the efficiency of operations in shifts of relaxed load situation. In order to keep operations efficient, a regular level of manpower has to be specified by the human planner as a positive estimate of the manpower demand over the shifts. Now it is subject of the automated planning procedure to place tasks such that the manpower demand is drawn as close as possible towards the level of regular manpower given. In this way the manpower consumption is balanced by ensuring efficient operations over the shifts considered.

Figure 5.3 illustrates a solution of 17 tasks placed in 4 shifts. Within each shift, horizontal guillotine cuts indicate the number of gangs incorporated, i.e. 3 gangs for the first shift. Time-window or precedence constraints hinder a seamless integration of tasks, e.g. for the uppermost gang in shift number one idle-time has to be accepted in the middle of the shift's time-span. The highest vertical extension of tasks in a shift determines its manpower requirement for this shift. The dashed line represents the regular level of manpower prescribed as an optimistic estimate of the human planner.

An even better solution would aim at filling the gaps in shifts one, two and four. This will hopefully lead to a manpower level that is closer to the regular level of manpower than given in the solution depicted in Figure 5.3. Finally a transfer of transshipment volume or a change of productivity is needed in order to increase the manpower demand for shift number three. The losses accompanied with such a change may lead to the opportunity of a further reduction of the manpower demand for shifts number one, two or four.

## 5.2 An Integrated Transshipment Model

First we model problem resources to be considered, i.e. time, tasks, manpower, storage areas and transfer points, and finally the productivity based telemetry in order to cover transport costs. Then, the variables of the model are described, before we turn to detailed considerations of the several classes of constraints involved. We differentiate temporal constraints from gang-oriented, manpower, location and inventory constraints. Finally we formalize the objective function proposed in Chapter 5.1.3.

In order to separate input data from variables, we denote the former with capital letters and the latter with lowercase letters. Central figures are the number of time ticks $T$, the set of tasks $A$ and the set of storage areas $F$. Constraint sets are typically stated by using dynamically generated subsets of $T, A$ and $F$. Subsets are expressed by $S$ : *condition*, denoting a subset of set $S$ for which *condition* holds.

### 5.2.1 Problem Resources

**Tasks.** For task $j$ a certain number of vehicles $L_j$ are to be moved in a time interval specified by its earliest starting time $EST_j$ and its latest finishing time $LFT_j$. Vehicles of a task are either to be stored ($Y_j = \texttt{S}$) or retrieved ($Y_j = \texttt{R}$). In case of storage $Q_j$ denotes the given origin, whereas in case of retrieval $Z_j$ prescribes the destination. Clearly, the destination of a storage task is subject to search and is therefore modeled as decision variable. The role of the origin of retrieval tasks is not that obvious:

- In case of a coupled transshipment, i.e. storage and retrieval task of a certain number of vehicles fall into the same planning horizon, the origin of the retrieval depends on the destination of the storage task.
- In case of an uncoupled retrieval task, the origin is specified by $Q_j$.

$A$ set of tasks $j \in A$

$L_j$ number of vehicles relocated by task $j$

$EST_j$ earliest starting time of task $j$

$LFT_j$ latest finishing time of task $j$

$Y_j \in \{\texttt{S}, \texttt{R}\}$ denotes type (storage, retrieval) of task $j$

$Q_j$ origin of task $j$ for $j \in A : Y_j = \texttt{S}$

$Z_j$ destination of task $j$ for $j \in A : Y_j = \texttt{R}$

$V_j$ predecessor task of task $j$ for $j \in A : Y_j = \texttt{R}, V_j = \emptyset$ otherwise

If tasks $i$ and $j$ are coupled so that $i$ precedes $j$, then $V_j = i$, $Y_i = \texttt{S}$ and $Y_j = \texttt{R}$. Furthermore, for both tasks $i$ and $j$: $L_j = L_i$ and $Q_j = Z_i$.

**Time.** The terminal operations are performed during separated shifts. There are two shifts per day, and each shift comprises 7.5 working hours. Discrete time steps model the flow of time. Since a resolution of 1/2 hour is used, each shift consists of $t_s = 15$ time ticks $t$. Although not limited by the model, let us consider that at most 19 consecutive shifts are planned simultaneously. Since shift boundaries are not stated explicitly, a total of $T = 19 \times 15 = 285$ ticks are considered.

$t$ time ticks for the entire planning horizon, $t = 0, \ldots, T$

$t_S$ the number of time ticks per shift is set to a prescribed value

$u$ the shift number of time tick $t$ can be calculated by $u(t) := \left\lfloor \frac{t}{t_S} \right\rfloor$

**Manpower.** The model provides a regular number of drivers $R_u$ on a per shift basis, which is chosen close to, but typically below, the expected demand. Since performing a task requires an administrative overhead, its minimal driver utilization is restricted to a useful number of drivers $R^{\min}$. We can suspect bottlenecks in the traffic system, i.e. bridges crossing rail connections as shown in Figure 4.1. Therefore we suppose a decreasing benefit of engaging additional drivers beyond a certain number. Hence we provide a limitation $R^{\max}$ on the number of drivers per task in the model. In practice, between 5 and 50 drivers perform a task.

$R_u$ regular manpower (number of drivers) employed in shift $u$

$R^{\min}$ minimum number of drivers required performing a task

$R^{\max}$ maximum number of drivers allowed performing a task

**Locations.** For internal locations, indicated by a type descriptor $H = \texttt{I}$, a capacity $K$ and inventory levels $B$ are considered. External locations with $H = \texttt{E}$ serve as transfer points and consequently no capacities or inventory levels are maintained, cf. Figure 5.4. Car carrier operations have to be performed under spatially narrow conditions; therefore a maximal number of simultaneously operating drivers $M$ is specified for a location.

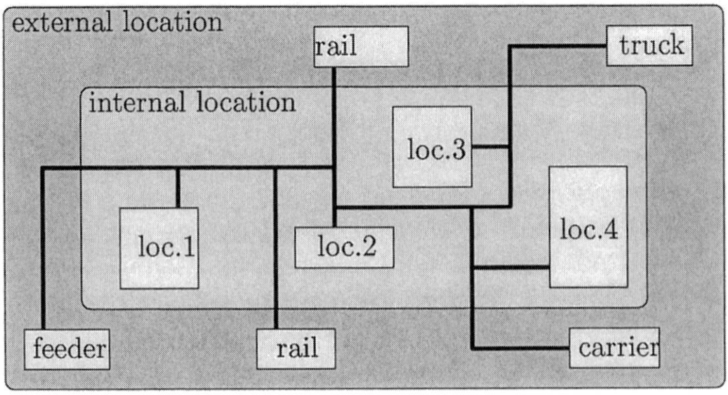


*Figure 5.4.* Illustration of terminal as considered in the planning and scheduling module. Internal locations represent storage areas of certain capacity, whereas external locations merely represent transshipment points. Locations are connected by a system of travel ways.

$F$ set of parking lots $i \in F$

$H_i$ $\in \{\mathtt{I}, \mathtt{E}\}$ describes type (internal, external) of location $i$

$K_i$ capacity of internal location $i$

$B_i$ initial inventory level of vehicles of internal location $i$

$M_i$ maximal number of drivers working simultaneously in location $i$

**Distances.** Productivity $\varphi(i_1, i_2)$ between location $i_1$ and $i_2$ determines the number of vehicle movements between $i_1$ and $i_2$ one driver can perform during a time tick. Analogously, the production coefficient $\varphi^{-1}$ gives the time needed to perform a single vehicle movement (cycle).

$\varphi(i_1, i_2)$ productivity between location $i_1$ and location $i_2$ with $i_1, i_2 \in F$

### 5.2.2 Decision Variables

**Storage Areas.** Only the destination of storage tasks $z_j$ can be subjected to a search. In case of a coupled transshipment, the origin of retrieval $q_j$ equals the destination of its logical predecessor, i.e. $q_j = z_{V_j}$. For this reason origins are also modeled as (dependent) variables.

$z_j$ $\in F$ destination location of task $j$

$q_j$ origin location of task $j$, if $V_j \neq 0$

**Manpower demand.** Since the number of gangs and their size differ from shift to shift, we do not model gangs explicitly. Instead the number of drivers utilized in a gang is stored as attribute $p_j$ of its tasks. Thus, all tasks a gang performs during one shift have the same number of drivers assigned to them. Since different gangs can have the same manpower demand, $p_j$ does not suffice to uniquely determine a gang.

Therefore we model a gang as a chain of predecessor relations $n_j$ among tasks. The first task $j$ in the chain with $n_j = \emptyset$ stands proxy for the implementation of a gang with $p_j$ drivers assigned to it.

$p_j$ $[R_{\min}, \ldots, R_{\max}]$ number of drivers employed for task $j$

$n_j$ $\in A$ predecessor task of $j$ in the same gang chain

**Starting and Completion Times.** We can derive starting times $s_j$ from a gang chain by assuming left shifted scheduling at the earliest possible starting time. Similarly the completion time $c_j$ of a task is fully determined, see Equation (5.19).

$s_j$ $[1, \ldots, T]$ starting time of task $j$

$c_j$ $[1, \ldots, T]$ completion time of task $j$

**Inventory Control.** Inventory levels are maintained for each internal location and every time tick. External locations are not considered here, because they are customer-owned and merely serve as transfer points for storage and retrieval tasks. Clearly the modifications of inventory levels depend on the starting- and completion time of the tasks involved.

$l_{ti}$ inventory of location $i$ at time $t$

Independent variables of central denotation are the destination location in case of storage $z_j$, the number of driver performing a task $p_j$ and the implicit assignment of a task to a gang $n_j$. These figure fully determine a solution to the problem, all other variables are derived by means of constraints to be considered.

### 5.2.3 Constraints

**Temporal Constraints.** Equations (5.1) ensure that the starting- and completion time of task $j$ fall into the same shift, i.e. tasks cannot be processed across shift boundaries. Time windows of tasks are taken into account by Equations (5.2) and (5.3). In case of coupled tasks precedence relations are considered by Equations (5.4) ensuring that tasks do not overlap in time. Note that time is considered in intervals instead of points in time, thus we have used the "strictly greater" operator in Equations (5.4).

$$\begin{aligned}
u(s_j) &= u(c_j), & \forall j \in A && (5.1)\\
s_j &\geq EST_j, & \forall j \in A && (5.2)\\
c_j &\leq LFT_j, & \forall j \in A && (5.3)\\
s_j &> c_{V_j}, & \forall j \in A : V_j \neq \emptyset && (5.4)
\end{aligned}$$

**Gang Constraints.** Equations (5.5) ensure that all tasks of a gang fall into the same shift. Equations (5.6) avoid the splitting of gangs by ensuring that no two tasks share the same predecessor. Finally, Equations (5.7) enforce that tasks belonging to the same gang have the same number of drivers assigned to them. Hence we can interchangeably use the terms *gang* and *task* in the context of manpower requirements. Similar to Equations (5.4), Equations (5.8) ensure that tasks assigned to one gang do not overlap in time.

$$\begin{aligned}
u(s_j) &= u(s_{n_j}), & \forall j \in A : n_j \neq \emptyset && (5.5)\\
n_j &\neq n_k, & \forall j, k \in A : j \neq k \wedge n_j \neq \emptyset \wedge n_k \neq \emptyset && (5.6)\\
p_j &= p_{n_j}, & \forall j \in A : n_j \neq \emptyset && (5.7)\\
s_j &> c_{n_j} & \forall j \in A : n_j \neq \emptyset && (5.8)
\end{aligned}$$

**Manpower Constraints.** Equations (5.9) and (5.10) restrict the number of drivers assigned to a gang. Equations (5.11) implement a more intricate constraint on the grouping of drivers into gangs. Drivers may hinder each other while working at the same location (even if they perform different tasks). Thus, the maximum number of drivers simultaneously allowed at location $i$ can be restricted.

$$\begin{aligned}
p_j &\geq R^{\min}, & \forall j \in A && (5.9)\\
p_j &\leq R^{\max}, & \forall j \in A && (5.10)\\
\sum_{j \in A : (q_i = i \vee z_i = i) \wedge s_j \leq t \leq c_j} p_j &\leq M_i, & \forall i \in F, \forall t = 0, \ldots T && (5.11)
\end{aligned}$$

**Location Constraints.** Equations (5.12) assign the prescribed origin for storage tasks as well as for uncoupled retrieval tasks. The prescribed destination of retrieval tasks is assigned by Equations (5.13). In case of coupled tasks, a predecessor of the retrieval task exists. Equations (5.14) state that the destination of storage equals the origin of retrieval. Furthermore, Equations (5.15) restrict the destination of storage tasks to internal locations. In this way the storage into transfer points is prevented.

$$
\begin{aligned}
q_j &= Q_j, & \forall j \in A : Y_j = \mathtt{S} \vee (Y_j = \mathtt{R} \wedge V_j = \emptyset) && (5.12)\\
z_j &= Z_j, & \forall j \in A : Y_j = \mathtt{R} && (5.13)\\
q_j &= z_{V_j}, & \forall j \in A : Y_j = \mathtt{R} \wedge V_j \neq \emptyset && (5.14)\\
H_{z_j} &= \mathtt{I}, & \forall j \in A : Y_j = \mathtt{S} && (5.15)
\end{aligned}
$$

**Inventory Constraints.** Equations (5.16) assign an initial inventory level at $t = 0$ to all internal storage areas $i$. The set of dynamic inventory balance equations (5.17) maintains the inventory level from $t = 1$ to $T$. Only internal areas $i, (H_i = \mathtt{I})$ are taken into account, such that the $L_j$ vehicles of storage task $j, (z_j = i)$ are added at $t, (s_j = t)$. Conversely, the $L_j$ vehicles of retrieval task $j, (q_j = i)$ are subtracted at $t, (c_j = t)$. This formulation considers each task $j$ twice by removing vehicles as early as possible from $q_j$ and by adding them as late as possible to $z_j$. In this way buffer times are provided in order to avoid traffic jams. Equations (5.18) keep the inventory level within the feasible domain.

$$
\begin{aligned}
l_{0,i} &= B_i, & \forall i \in F : H_i = \mathtt{I} && (5.16)\\
l_{t,i} &= l_{t-1,i} - \sum_{\substack{j \in A: \\ q_j = i \wedge c_j = t}} L_j + \sum_{\substack{j \in A: \\ z_j = i \wedge s_j = t}} L_j, & &\\
& & \forall i \in F : H_i = \mathtt{I}, \forall t = 1, \ldots T && (5.17)\\
l_{t,i} &\leq K_i, & \forall i \in F : H_i = \mathtt{I}, \forall t = 1, \ldots T && (5.18)
\end{aligned}
$$

**Completion Time.** Equations (5.19) determine the completion time of a task dependent on its starting time and duration. The duration of a task depends on the number of vehicle relocation cycles required and the duration of an individual cycle, resulting in a non-linear constraint. The vehicle volume $L_j$ and number of drivers $p_j$ determine the number of cycles. A non-integer value of $L_j/p_j$ indicates that only a subset of drivers can be used in the last cycle. The rounding to the next larger integer implements that the remaining drivers may have to wait for their driving colleagues during the last cycle.

The duration of an individual cycle is given by the coefficient $\varphi^{-1}$, which depends on the productivity measure between the associated storage areas. Fractional durations are rounded up to the next time tick.

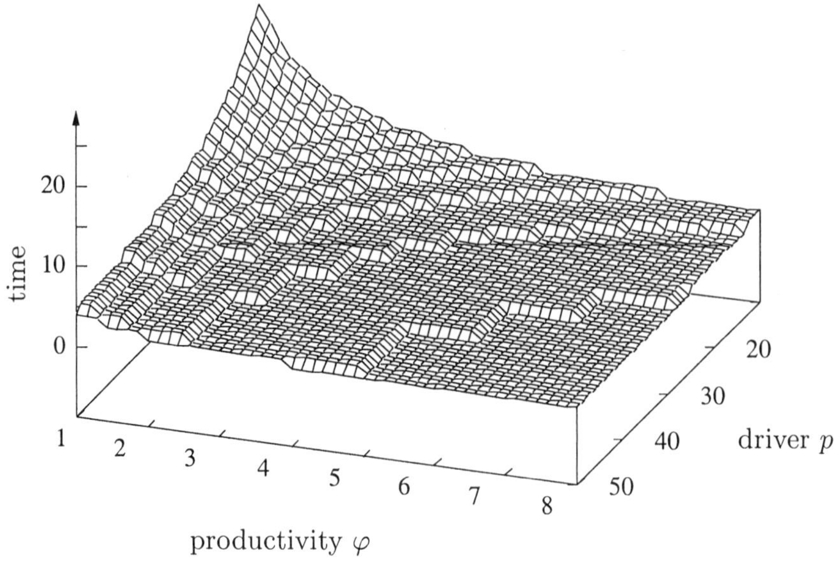


*Figure 5.5.* Example of the impact on a task's processing time by the productivity and the number of driver involved.

$$c_j = s_j + \left\lceil \left\lceil \frac{L_j}{p_j} \right\rceil \cdot \varphi^{-1}(q_j, z_j) \right\rceil, \qquad \forall j \in A \tag{5.19}$$

Figure 5.5 illustrates the dependencies expressed by Equations (5.19) for a task of $L_j = 200$ vehicles. The duration of processing decreases by either increasing the number of drivers $p_j$ or by increasing the productivity $\varphi$.

The non-linearity of the constraint can be easily recognized from Figure 5.5, where the duration of processing increases non-linearly with a decreasing number of drivers $p_j$ or/and a decreasing productivity $q_j, z_j$ of task $j$ involved.

The integer conditions with respect to the vehicle relocation cycles performed as well as with respect to time intervals covered lead to the stairways-like function with plateaus of identical performance. Whenever there is a choice, efficient points of a plateau should be implemented. For instance, a productivity of 5 and 40 drivers denote an efficient point. Neither a further increase of productivity nor an increase of the number of driver involved can lead to an additional shortage of the processing time.

### 5.2.4 Objective Function

Since the tasks are externally defined and therefore prescribed for a problem instance, solely the manpower requirements can be subject to optimization. A minimization of the total number of drivers summed up over the shifts considered will probably lead to a cost minimal solution.

However, an uneven usage of manpower will not comply with the quality issues of operations. From the viewpoint of quality management a leveling of the manpower demand over the shifts considered is preferable.

We pursue a combination of both goals by minimizing the deviation of manpower demand from a prescribed (typically optimistic) regular level given by $R_k$ for shift $k$.

$$P_k = \sum_{j \in A : u(s_j) = k \wedge n_j = \emptyset} p_j \tag{5.20}$$

Equation (5.20) determines the driver demand $P_k$ for shift $k$. This figure is easily calculated by summing up $p_j$ over a representative task $j$ for each gang ($n_j = \emptyset$ considers the first task of a gang only) that is processed in shift $k$.

$$\min f(z, p, n) = \sum_{k=1}^{u(T)} (R_k - P_k)^2 \tag{5.21}$$

By taking Equation (5.1)-(5.19) into account, a solution to the problem is fully determined by an assignment of the decision variables $z_j$, $p_j$ and $n_j$ for all tasks $j \in A$. Equation (5.21) minimizes the squared deviation of $P_k$ from the regular level $R_k$ over the shifts considered. In this way a unit of a large deviation is penalized more highly compared to a unit of a small deviation.

By comparison with Figure 5.3, we immediately grasp that summing up the driver demand of an arbitrary task for each gang implemented in a single shift $k$ results in the total driver demand $P_k$ considered. The minimization of the squared deviation to the regular demand $R_k$ (given as dashed horizontal line in the figure) is a meaningful objective to balance the manpower demand.

Since the demand $P_k$ is drawn towards $R_k$, a reduction of the sum of the manpower demand is pursued whenever $R_k$ is low (enough). In the event that $R_k$ is set too low the focus of the optimization gets lost, because the deviation to $P_k$ is not meaningful anymore. With $R_k$ set properly, however, the objective function formulation aims at leveling demand peaks while increasing the overall productivity of the terminal at the same time.

## 5.3 Problem Separation

### 5.3.1 Implications of the Integral Model

For the matter of problem separation Schneeweiss (1999) differentiates between an organizational hierarchy due to information asymmetries and a constructional hierarchy, which reduces conceptual and/or computational complexity. In the following we show that both issues exist for the integral model presented, which calls for a separation of the problem.

**Constructional Hierarchy.** The model defined above comprises a large number of integral variables and at least one set of non-linear constraints of central meaning. Characteristics of the problem are related to the

- general assignment problem: This problem considers the assignment of items to sites. Although efficient algorithms exist for special cases of the problem, the more general formulation, e.g. allowing the assigning of multiple resources remains as a challenge (Gavish and Pirkul, 1991). Variants of the problem differ with respect to the number and type of constraints involved (Laguna et al., 1995).

- transshipment problem: This variant of the famous transportation problem was introduced by Orden in 1956 (Williams, 1999). Although the transportation problem decides on flows between various sources and sinks, this formulation can be used to model assignments between sources and sinks as well. In the case of transshipment, beside the explicit distinction of sources and sinks, nodes can act as sources and sinks at the same time.

- dynamic lot-sizing problem: Dynamic lot-sizing decides on the periods of producing a good in a certain quantity under the assumption of a costly storage of goods beyond the period of production (Haase, 1993). The multi-period formulation of the problem necessitates dynamic inventory balancing equations in order to couple the inventory of the periods considered.

- resource constrained project scheduling : This problem is concerned with the assignment of resources to tasks over time (Kolisch, 1995). The scheduling problem becomes difficult because of logical precedence relations between tasks encountered as well as earliest and latest finishing times considered for tasks. A recent work considers the allocation of resources over time, which allows the modeling of the intermediate storage of goods (Schwindt, 2002).

Since for all of the above problems the generation of an optimal solution remains as a computational challenge, we cannot expect to solve the integral problem sufficiently fast (if at all).

Also the problem size may prohibit solving a problem instance to optimality. Bear in mind, that more than 1,000,000 vehicles are transshipped via large terminals per year. This figure has to be doubled, because each vehicle is moved twice. Approximately 6,000 vehicles are processed per day. By assuming a mean task volume of 200 vehicles, 30 tasks have to be processed. By taking a planning horizon of 10 days only, 600 tasks have to be assigned to periods, reasonable storage areas have to be found among dozens of locations, and finally, within each shift, the tasks to be distributed to several gangs of varying size.

**Organizational Hierarchy.** Due to the stochastics of the available data, a solution procedure will be applied iteratively in the framework of a decision support system. A human planner is able to modify critical input data interactively. Therefore the process of evolving a final solution typically requires a number of successive optimization cycles alternated with data modifications performed by a human planner. This process requires a solution procedure, which allows for problem refinement during successive optimization cycles, and generates solutions as quickly as demanded by interactivity.

To cope with the argument of successive data refinement, in the following we separate the integral model into a mid-term planning model and a short-term scheduling model. Besides the obvious benefit of reducing problem complexity, organizational advances can be stated:

- Planning and scheduling is applied on the basis of a rolling time horizon, cf. Figure 5.6. As time passes by, data about future terminal operations is becoming more accurate. The tasks of a shift are replanned with varying data, before the shift is eventually deployed and finally implemented. The planning of far-termed shifts can be done approximately on a mid-term level only. The relatively bad data quality concerning far-termed shifts does not justify a detailed scheduling from the first time of consideration on.

- A human planner can decide on the number of shifts the detailed scheduling actually covers. In the extreme, one may even omit any detailed scheduling while merely relying on estimates obtained from mid-term planning. Scheduling can then be integrated into the optimization course successively in later cycles. Thus, the separation of mid-term planning and short-term scheduling takes up the existence

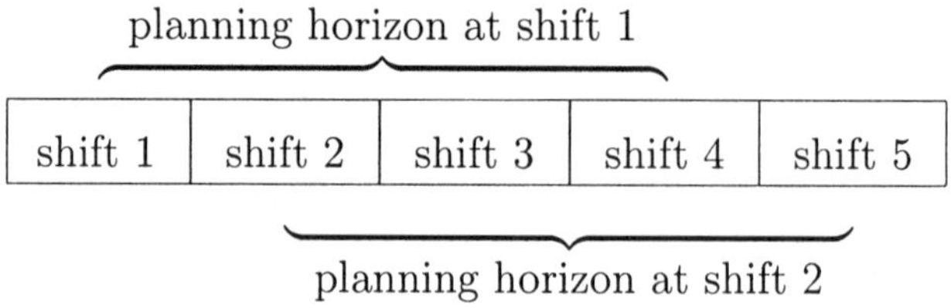


*Figure 5.6.* Scheme of planning on a rolling time horizon.

of information asymmetries for temporal as well as for organizational reasons.

In terms of Schneeweiss (1999), the concept of problem separation supports both, an organizational hierarchy due to information asymmetries , and a constructional hierarchy, which reduces conceptual and/or computational complexity. Although for our purpose the reduction of computational complexity is aimed at in the first case, also information asymmetries exist due to the stochastics of data. Obviously, data about tasks to process in the future are less specific than task data belonging to the current shift of execution. We benefit from a mid-term planning based on relatively rough data, which can be iteratively refined by short-term scheduling in later stages of the planning procedure.

Such an approach has to integrate mid-term capacity planning and short-term scheduling, see Nobert and Roy (1998), for an application at an air cargo terminal. The authors describe the capacity planning as a problem of determining the manpower requirements in a given planning period. The scheduling problem consists of designing efficient work schedules that satisfy the manpower requirements and comply with regulations that apply.

The problem separation is presented in detail in Section 5.3.2.Furthermore we consider heuristics, because the sub-problems generated by the problem separation are still too complex to be solved exactly. A reasonable way of solving the mid-term planning problem is given in detail in Chapter 6. Finally the development of a reasonable heuristic for the short-term scheduling problem is subject of Chapter 7.

### 5.3.2 Hierarchical Problem Separation

We separate the integral model into a mid-term oriented allocation of storage space, i.e. an assignment of tasks to periods and storage space to tasks (top-level), and a short-term oriented personnel deployment, i.e. scheduling of tasks into gangs of a shift (base-level).

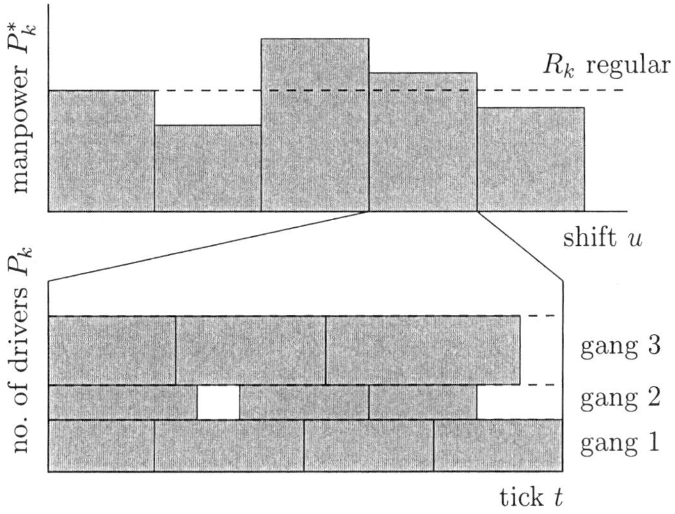


*Figure 5.7.* Scheme of the hierarchical separation into a top-level planning model and a base-level scheduling model to be solved for each shift separately. The top-level model considers a manpower aggregate in terms of "driving hours" while the base-level model considers the number of drivers directly.

**Top-Level Model.** The model decides upon the processing shift $u(s_j)$ and the storage area $z_j$. Therefore, constraints (5.1)–(5.4) are relaxed by considering a single tick per shift only, i.e. $s_j = c_j = u(s_j) = u(c_j)$. Equations (5.5)–(5.11) are no longer relevant, because neither gangs nor drivers are considered in the top-level model. Decisions regarding the choice of storage locations must satisfy Equations (5.12)–(5.15). All decisions to be taken are linked by a shift-oriented relaxation of the inventory constraints (5.16)–(5.18). Since the duration of tasks is relaxed, i.e. $s_j = c_j$, different to the integral model a storage area can be reused immediately at the time it is emptied.

$$P_k^* = \left\lceil \frac{1}{t_S} \sum_{j \in A : u(s_j) = k} L_j \cdot \varphi^{-1}(q_j, z_j) \right\rceil \tag{5.22}$$

Since $p_j$ is not defined in the top-level model, Equation (5.20) is not applicable. Therefore we determine $P_k^*$ in Equation (5.22) to estimate the manpower demand for shift $k$ by dividing the aggregate "driving hours" by the number of ticks per shift $t_S$. Since $P_k^*$ is a lower bound on the actual driver demand $P_k$, we can still use Equation (5.21) as the objective function with the only difference of using $P_k^*$ instead of $P_k$.

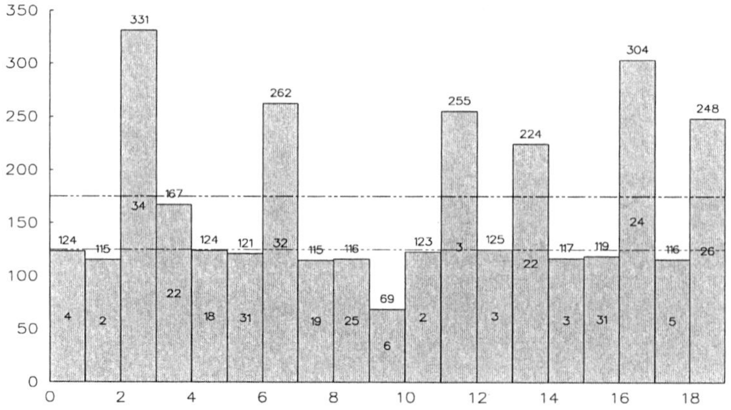


(a) initial solution obtained from a one-pass heuristic

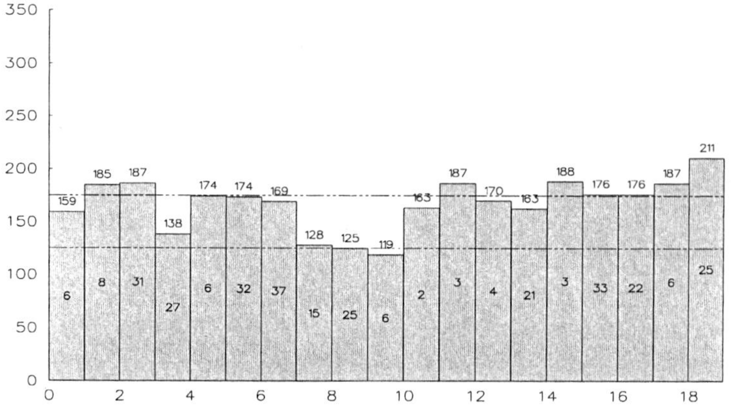


(b) improvements are obtained by further modifications

*Figure 5.8.* Example of a solution to the mid-term model for operations planning.

Before we proceed by deriving a mathematical model for the base-level problem, we sketch the mid-term model by an example. Figure 5.8 depicts two solutions for 19 shifts with 312 tasks in different stages of the optimization course. The x-axis denotes the shift $k$, whereas the y-axis depicts the manpower capacity. The number inside a shift-column reports the number of tasks processed in that shift. The number on top of a shift-column reports its anticipated driver demand $P_k^*$.

Two horizontal lines are shown in the figure. The above line refers to the mean manpower demand over the shifts for solution (a). Note, that in solution (b) almost all shifts show a smaller manpower demand than

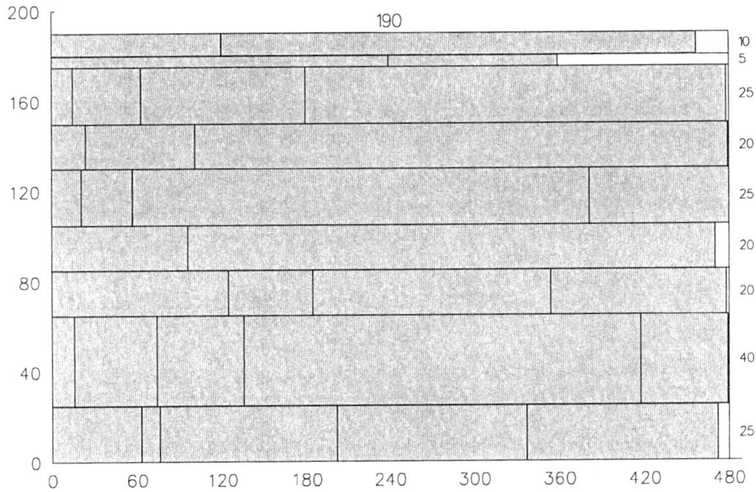


*Figure 5.9.* Example of short-term operations scheduling.

the mean demand of (a). Obviously, significant efficiency improvements with respect to the manpower utilization have been gained. The lower horizontal line refers to the regular manpower demand $R_k$.

The extremely high manpower demand of 331 drivers for shift 2 in solution (a) is decreased by moving tasks into shifts 0 and 1 in solution (b). If time windows are narrow, a movement of tasks may not be feasible, as it is the case (data not shown) for shift 11 in the example. In order to gain improvements here, alternate location assignments are considered.

The choice of remote locations in shifts with a relaxed manpower demand is by no means a waste of capacity. Rather, distant storage areas are chosen in order to employ the number of regular drivers $R_k$. In this way central locations of potentially high productivity may be preserved for use in a forthcoming congested shift. Here, the driver demand for shifts 7,8 and 9 increases in (b), whereas the demand of shifts 10 and 11 significantly decreases.

**Base-Level Model.** At the base-level, operations scheduling can be carried out for each shift separately. Scheduling receives the locations $q_j$ and $z_j$ and the shift $u(s_j)$ as input data from the top-level. The temporal constraints (5.2)–(5.4) apply in their original setting. Furthermore, gang related constraints (5.5)–(5.7) and manpower related constraints (5.9)–

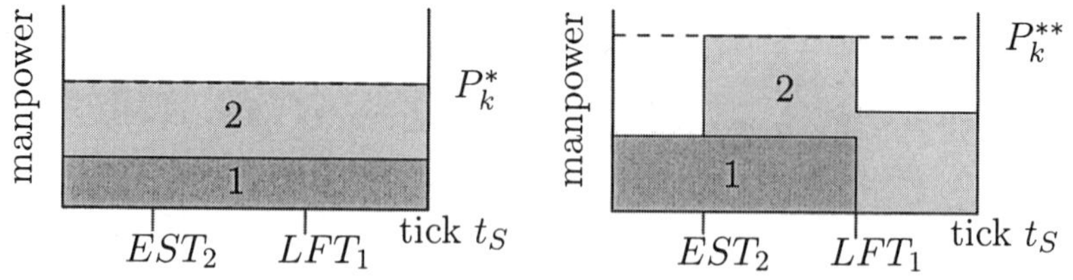


*Figure 5.10.* Scheduling is (partly) anticipated by a deterministic simulation in the top-level model. By considering $EST_2$ and $LFT_1$ at the level of time ticks for shift $k$, the actual manpower demand can be much better estimated.

(5.11) apply. Instead of controlling the inventory in the detailed model, precedence constraints as expressed by Equations (5.4) are inserted for all tasks $j_1, j_2 \in A$ if $j_2$ re-uses a certain storage area directly after it has been emptied by $j_1$.

Since for the base-level model Equations (5.19) apply, Equation (5.20) can be used to determine the actual manpower demand $P_k$, i.e. the number of drivers required for the shift. The goal of the base-level problem is to draw $P_k$ as close as possible towards $P_k^*$. Minimizing $P_k$ can make this goal operational.

Although, at a first glance, operations scheduling shows apparent similarities with multi mode project scheduling (Brucker et al., 1999), it differs in the objective function pursued and in the minor role of precedence relations to be considered. The introduction of gang constraints requires the consideration of two successive problems. At the upper level, tasks are assigned to gangs, whereas at the lower level the manpower-minimal order of tasks is determined for each gang separately.

The assignment of tasks to gangs does not completely specify a solution, since different task sequences within a gang may still be feasible. Thus, as a sub-problem the manpower-minimal task sequence has to be calculated. Figure 5.9 presents a fairly good solution for shift no. 2 of Figure 5.8(b) with 31 tasks and an approximated manpower demand of 187 drivers. The tasks are depicted over the 480 minutes of a shift (x-axis) requiring a total of 190 drivers organized in 9 gangs.

**Base-Level Anticipation.** The validity of the hierarchical separation depends on how well $P_k^*$ approximates $P_k$. If there is a weak correlation only, the top-level model will take unfavorable decisions with respect to the base-level model. According to Schneeweiss (1999), there should be an anticipation of the base level. We follow Schneeweiss by generating an approximate schedule already at the top level, see also Schneeweiss and Zimmer (2004).

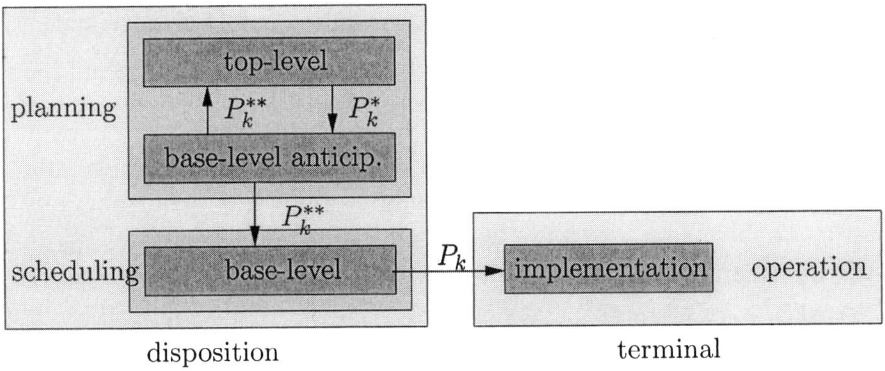


*Figure 5.11.* Integration of the hierarchical separation of the solution procedure. The anticipatory simulation is integrated in the top-level procedure.

In so doing a time resolution at the tick level is taken into account, which allows a partial support of Equations (5.2)–(5.4). The starting and completion times are set to the prescribed earliest starting times and latest finishing times, i.e. $s_j = EST_j$ and $c_j = LFT_j$. This consideration of task durations at the level of time ticks improves $P_k^*$ to $P_k^{**}$.

On the left of Figure 5.10 manpower capacity is treated on a per-shift basis in accordance with Equation (5.22) of the top-level model. The right side shows the estimated number of drivers $P_k^{**}$ obtained by the anticipated schedule construction, which will be considerably closer to $P_k$ compared to $P_k^*$.

Figure 5.11 shows the integration of the anticipatory simulation in the two-level solution procedure. The top-level produces $P_k^*$ which is refined by the base-level anticipation to $P_k^{**}$. This cycle can be run several times before the solution (for which $P_k^{**}$ has been determined) is irrevocably passed on to the base-level. On this basis, $P_k$ is generated and finally implemented.

Since the top-level problem considers an aggregate of the actual manpower demand to be considered, we are able to run the top-level model without support from the underlying base-level model, while retaining a meaningful result for the problem instance under consideration. When the disposition applies the procedure on a rolling time horizon, usually many shifts are involved in a problem instance. Typically only a small subset of shifts in the near future has to be implemented. Only for these shifts the base-level model has to be considered.

## 5.4 Summary

In this chapter we have modeled vehicle terminal operations under the goal of meeting customer expectations. In particular, low damage rates are essential for successful business, which can be achieved by balancing the workload.

A balanced workload can be obtained first by shifting transport tasks forth and back, and second, by an appropriate allocation of storage space with respect to the transport distance covered. Moreover, safe and reliable operations are achieved by a deployment of personnel into gangs.

We have shown that the decisions to be taken highly interrelate with each other, leading to a complex optimization problem. Several resources enter the model as input data, namely, discrete time steps, transshipment tasks, storage areas and transfer points and finally a productivity based distance metric for the locations considered.

The resulting model depicts the apparent dependencies between the choice of the storage area and the mode of task processing, i.e. its utilization of the transport facility. In this way the model seems generally applicable to transshipment problems where the distance to be covered and the amount of transport facility needed interrelate.

The complexity of the model suggests a hierarchical separation into a mid-term consideration of terminal and manpower capacity and a short-term consideration of personnel deployment. Both sub-problems derived will be still too complex in order to be solved exactly. However, the separation may yield a reasonable problem difficulty, which allows the application of heuristics.

Whenever an integral model is hierarchically separated, an anticipation of base-level decisions has to be considered with respect to decisions taken at the top-level. Clearly, the ways of anticipating future decisions are problem specific. For our purpose we have integrated certain aspects of detailed scheduling into the period-oriented planning model.

# Chapter 6

# ALLOCATION OF STORAGE SPACE

**Abstract** This chapter addresses the allocation of storage capacity over time. Transshipment tasks compete for storage space in spatially distributed storage areas of finite capacity. Although the optimization model developed suggests considering the assignment of tasks individually, the Evolutionary Algorithm proposed evolves a capacity utilization strategy. This capacity utilization strategy then controls a construction heuristic, which assigns tasks to periods and allocates storage areas to tasks. A well adopted construction heuristic allows solving large instances of space allocation problems.

## 6.1 Space Allocation Problems

The space allocation is an important task in the mid-term planning of vehicle transshipment operations. Since vehicles cannot be stacked, the land use of terminals can be enormous. Thus, the transportation effort plays a major role for the efficiency of a terminal. Unfortunately, capacity constraints on the storage areas implemented in multi-period models are needed in order to depict transshipment processes reasonably well. Moreover, a large number of storage areas and an even larger number of transshipment tasks have to be considered. In the following we present a heuristic approach to cope with the problem at hand.

First, we give a brief overview on recent approaches to related problems. In Section 6.1.2 we describe our problem and present a mathematical formulation of the problem in analogy to Section 5.2. In Section 6.2 a procedure for the generation of test problems is proposed. In Section 6.3 we describe the construction heuristic controlled by the capacity utilization strategy. In Section 6.4 the adaptation of this strategy is subject to an Evolutionary Algorithm, for which we perform a thorough computational investigation.

### 6.1.1 Literature Overview

Space allocation problems consider the allocation of storage capacity to inventory with respect to transportation effort (Kusiak, 2000, 412-427). Whenever the deviation of the transshipment volume over time is rather small, the problem can be solved by modeling one period only. The assignment of storage areas to tasks can be done by a variant of the standard transportation problem, which minimizes the transportation effort between spatially distributed storage areas.

Already in 1956, Orden introduced the transshipment problem, which extends the model by the flow of commodities from origins to destinations via intermediate transshipment points (Williams, 1999). Nowadays, advanced multi-stage, multi-commodity distribution models consider sources and destinations interconnected by distribution centers and consolidation points (Fleischmann, 2005). These models, however, primarily address strategic decisions because of their simplifying assumptions.

Provided that storage in standardized racks, the space-allocation problem can be sourced out to a generously dimensioned automated storage/retrieval system (AS/RS)(Muralidharan et al., 1995). Research in this field is targeted at finding reasonable or even optimal policies. Whenever the availability of storage space is constrained, the temporal dependencies between transshipment tasks with regard to the storage space consumption have to be taken into account. This must already be included at the level of task planning and cannot be deferred to the level of process control at the AS/RS.

To support this goal, storage space allocation problems may be modeled as vehicle routing problems (Bramel and Simchi-Levi, 1997) with intermediate storage facilities (Angelelli and Speranza, 2002). This way of modeling maintains a focus on the utilization of the transport facility.

By focusing on precedence dependencies of transshipment tasks, we may model the problem as a resource constrained project scheduling problem where tasks can be performed in various modes (Brucker et al., 1999). Here, storage areas of finite capacity can be depicted as non-renewable resources (Neumann and Schwindt, 1999; Schwindt, 2002), which can be handled efficiently (Laborie, 2001). However, the problem complexity prohibits finding satisfying solutions to any reasonably sized problem at this detailed operational level.

Literature concerned with container transshipment stresses the particularities of equipment handling. The stacking of containers decreases the storage space needed and hence also decreases the mean transportation effort required. To the opposite handling work increases in case of stacking and therefore we observe a tradeoff between the consumption of storage space and the handling work required (Taleb-Ibrahimi and Castilho, 1993). The authors aim at calculating the minimum space required for a given transshipment rate. Furthermore, the minimal handling costs can be obtained for a given storage space. Both figures address strategical/tactical decisions only.

At an operational level Kozan and Preston (1999); Preston and Kozan (2001) suggest optimizing the allocation of storage space such that setup times (synonymous for handling work) and transport time become minimal. The authors use a Genetic Algorithm in order to generate a sequence of geo-coordinates, at which containers are to be placed. Stacking of containers is penalized by additional handling times, whereas unnecessary detouring during container placements are penalized by additional transport effort. The approach does not consider multiple periods of transshipment.

Time is incorporated either implicit by applying suitable rules for space allocation, i.e. based on the duration of stay of containers in a yard (Kim and Park, 2003). Bish et al. (2001) make timing explicit by means of a (single period) scheduling model. The authors minimize the time needed to unload a container ship under the constraints of limited availability of storage areas and transport vehicles. The model developed assigns container to storage areas and vehicles to containers. Handling work is not considered in this rather simple model.

Zhang et al. (2003) propose two successive MIP formulations to be solved for a multi-period problem. First, the workload is balanced over the storage areas available, before second, the transportation effort is minimized. Workload can be balanced between various periods by means of space allocation decisions. This multi-period assignment problem is applied on the basis of a rolling time horizon by currently adjusting the solution to forthcoming changing conditions.

In vehicle transshipment, handling work is assumed constant and therefore just transport effort is to be minimized in storage space allocation models. See Mattfeld (2003) for an earlier work on this subject. Here, the balancing of workload is of first importance. Different to container transshipment, where the buffering of containers at a marshaling area can be used to balance workload over time, in vehicle transshipment such buffer facilities do not exist. Because of reasons of reliability and safety, the number of movements of vehicles is to be kept at the absolute minimum. Furthermore, a balanced workload over the periods contributes to the safety of operations.

In the following we consider a capacitated multi-period space allocation problem where transportation tasks between two transfer points call for intermediate storage. Besides the choice of appropriate storage areas, time-windows allow for selecting periods of operation. Since transportation volumes fluctuate strongly over time, a balanced utilization of vehicle driving personnel is aimed at. The transportation effort constitutes an important cost driver; therefore efficient operations are also pursued.

The Evolutionary Algorithm proposed to solve this problem evolves capacity utilization strategies. The capacity utilization strategy controls a construction heuristic, which performs the actual plan generation. This algorithm is particularly well suited for the capacity constrained planning of space allocations on a rolling time horizon. The assignment of tasks to periods and the allocation of storage space have to be carried out for the periods of the planning horizon. Thereby the deviation of manpower demand over the periods is to be minimized while keeping the overall manpower consumption reasonably small.

### 6.1.2 Problem Modeling

We consider a transshipment terminal which consists of a network of spatially distributed (internal) storage areas of finite capacity interconnected by travel ways. This scope of the network is extended by (external) transfer points which depict the interfaces of the terminal to the outside world, i.e. berthing facilities, rail ramps, and dealer compounds. The manpower demand — standing proxy for other transportation costs — generated by a relocation of vehicles is determined as a function of the number of vehicles (volume) and the distance covered (Daganzo, 1999).

Services offered to customers comprise the transshipment of charges of vehicles from one transfer point to another. Customers do not necessarily insist on the transportation within a certain period, but grant time-windows for both relocation types, the storage into the transshipment terminal and its corresponding retrieval . In order to model that the storage and retrieval may fall asunder, we consider the transshipment of a charge of vehicles as a pair of storage and retrieval tasks coupled by a precedence constraint.

Whenever customer granted time-windows for a pair of tasks do not overlap, an intermediate storage becomes unavoidable. In these cases a storage area of sufficient capacity is chosen, such that the manpower demand for both logistic tasks, storage and forthcoming retrieval, is reasonably small. Whenever the time-windows for a pair of storage and retrieval tasks overlap, typically both tasks are assigned to the same period, i.e. the vehicles are driven directly. Since the consumption of storage space can be neglected in this case, storage capacity is not taken into account.

To summarize, the assignment of tasks to periods and the allocation of storage space has to be carried out for the periods of the planning horizon. Thereby the sum of the squared deviation of the manpower demand from a prescribed regular manpower level is to be minimized. The quadratic term is used to penalize large deviations strongly, while a reasonable small prescription of regular manpower in the objective function can be utilized to keep the overall manpower consumption reasonably small.

This problem comprises a large number of integer variables and can result in a highly constrained solution space. In the following, we describe the problem resources before going on to list the decision variables explicitly. Then we turn to a description of the objective function and the constraints involved.

### Resources Considered

#### discrete time model

$t$ periods of the planning horizon, $t = 0, \ldots, T$

#### storage areas and transfer points

$F$ set of locations

$H_i \in \{\mathtt{I}, \mathtt{E}\}$ denotes the type (internal, external) of location $i \in F$. Internal locations may be used for storage, while external locations (transfer points) merely act as origin or destination of vehicles

$K_i$ storage capacity of internal storage location $i \in F : H_i = \mathtt{I}$

$B_i$ initial inventory of internal storage location $i \in F : H_i = \mathtt{I}$

#### manpower capacity

$P_t$ level of regular manpower for period $t$ in terms of working-hours. This figure will be externally given by the terminal management as an optimistic estimate of the actual manpower demand. The model will minimize the squared deviation from this figure, such that large deviations are strongly punished.

#### productivity measure

$D_{i,k}$ production coefficient between location $i$ and location $k$ expressed working-hours required for a relocation of one vehicle.

#### transshipment tasks

$A$ set of all transshipment tasks to be performed

$Y_j$ $\in \{\mathtt{S}, \mathtt{R}\}$ denotes type (storage, retrieval) of task $j$

$L_j$ volume (number of vehicles) of task $j$

$Q_j$ origin of storage task prescribed for $j \in A : Y_j = \mathtt{S}$

$Z_j$ destination of retrieval task prescribed for $j \in A : Y_j = \mathtt{R}$

$V_j$ preceding storage task of retrieval task $j$, $V_j = \emptyset$ if not existing

$N_j$ succeeding retrieval task of storage task $j$, $N_j = \emptyset$ if not existing

$EST_j$ period of earliest permissible shipment of task $j$

$LFT_j$ period of latest permissible shipment of task $j$

**Variables**

We distinguish two types of decision variables, namely the assignment of tasks to periods $s_j$ and the allocation of storage space $z_j$ for storage tasks. Since we decide on the destination of storage tasks, the origin of the corresponding retrieval tasks $q_j$ is modeled as a dependent variable. A task $j \in A$ is fully determined by an assignment to $(s_j, q_j, z_j)$. Other variables have to be considered as dependent variables in order to implement appropriate constraints on the manpower demand $(p_t)$ and the storage capacity $(l_{t,i})$.

**task-to-period assignment**

$s_j$ $\in \{1 \ldots T\}$ period of shipment of task $j$

**storage area allocation**

$z_j$ $\in F$ destination location of storage task $j$

**retrieval location allocation**

$q_j$ $\in F$ origin of retrieval task $j$

**manpower planning**

$p_t$ actual manpower demand in working-hours for period $t$.

**inventory holding**

$l_{t,i}$ inventory of location $i$ in period $t$

### Objective Function and Constraints

$$\min \sum_{t=1}^{T} (P_t - p_t)^2 \tag{6.1}$$

$$\sum_{\substack{j \in A: \\ s_j = t}} L_j \, D_{q_j, z_j} = p_t, \qquad \forall t = 1, \ldots, T \tag{6.2}$$

$$EST_j \leq s_j \leq LFT_j, \qquad \forall j \in A \tag{6.3}$$

$$s_{V_j} \leq s_j, \qquad \forall j \in A : V_j \neq 0 \tag{6.4}$$

$$1 \leq s_j \leq T, \qquad \forall j \in A_j \tag{6.5}$$

$$q_j = Q_j, \qquad \forall j \in A : Y_j = \mathtt{S} \tag{6.6}$$

$$z_j = Z_j, \qquad \forall j \in A : Y_j = \mathtt{R} \tag{6.7}$$

$$H_{z_j} = \mathtt{W}, \qquad \forall j \in A : Y_j = \mathtt{S} \tag{6.8}$$

$$q_j = z_{V_j}, \qquad \forall j \in A : V_j \neq 0 \tag{6.9}$$

$$l_{0,i} = B_i, \qquad \forall i \in F : H_i = \mathtt{I} \tag{6.10}$$

$$l_{t-1,i} + \sum_{\substack{j \in A: \\ s_j = t \wedge z_j = i}} L_j = l_{t,i} + \sum_{\substack{j \in A: \\ s_j = t \wedge q_j = i}} L_j$$

$$\forall i \in F : H_i = \mathtt{I}, \forall t = 1, \ldots, T \tag{6.11}$$

$$0 \leq l_{t,i} \leq K_i, \qquad \forall i \in F : H_i = \mathtt{I}, \forall t = 1, \ldots, T \tag{6.12}$$

The objective function (6.1) minimizes the squared deviation of the manpower demand $p_t$ from the regular demand $P_t$. In this way, the deviation from the regular manpower demand can be kept reasonably small while minimization of the sum of manpower demand over the periods considered is pursed. In (6.2) $p_t$ is determined as the product of volume $L_j$ and production coefficient $D_{q_j,z_j}$ with regard to the tasks $j \in A : s_j = t$.

The constraints (6.3)-(6.5) refer to **temporal dependencies**. Constraints (6.3) keep the time-windows of tasks, constraints (6.4) ensure that precedence relations among coupled tasks are kept, i.e. retrieval task $j$ cannot be processed before its coupled storage task $V_j$. Finally, constraints (6.5) restrict the period of processing to the domain of planning periods considered.

The constraints (6.6)-(6.9) address **location requirements**. For storage tasks the origin is prescribed by (6.6) and for retrieval tasks the destination is externally given by (6.7). Constraints (6.8) ensure that space allocations are performed for internal locations only, whereas constraints (6.9) couple the storage locations of retrieval task $j$ and its predecessor task $V_j$.

Sets (6.10)-(6.12) refer to **inventory constraints**. Initial inventory levels are prescribed by (6.10). The dynamic inventory balance equations (6.11) necessitate that for every storage area $i$ and every period $t$ the inventory level of $t-1$ plus the volume stored in $t$ will be equal to the inventory level of $t$ diminished by the number of vehicles retrieved in period $t$. Finally (6.12) ensure that inventory levels do not exceed the area capacities given.

## 6.2 Problem Generation

In this section we present a way of producing solvable test problems for which at least one feasible solution exists. Finally we describe the parameterization of the set of 4500 test problems used later on.

### 6.2.1 Producing Solvable Test Problems

The generation of solvable test problems is a challenge. We commence by firstly generate a model of a terminal, and then proceed with the stepwise generation of transshipment tasks.

**Terminal Generation.** The terminal consists of internal storage areas $i \in F : H_i = \mathtt{I}$ and external transfer points $i \in F : H_i = \mathtt{E}$. The storage capacity $K_i$ of internal location $i$ is drawn from a uniform distribution, and the total capacity of the storage system is summed up to $\hat{K} = \sum_{i \in F:H_i=\mathtt{I}} K_i$.

The coordinates of the storage areas are distributed to a standard normal, while the coordinates of transfer points are generated from a uniform distribution in $[-3, 3]$. Since a variable drawn from the standard normal distribution falls into this interval with probability .997, storage areas tend to be located in the center of the transfer points.

The distances observed are taken as production coefficient $D_{i,k}$, i.e. the working-hours needed to transport one vehicle from location $i$ to location $k$. Figure 6.1 provides a bird's eye view to the locations and distances of the simulated terminal.

**Generating Storage Tasks.** Tasks are produced by simulating a transshipment scenario for a number of periods with respect to already existing terminal data. In this simulation, storage tasks are assigned to periods with increasing time. The inter-arrival time of storage tasks is consecutively drawn from an exponential distribution. The corre-

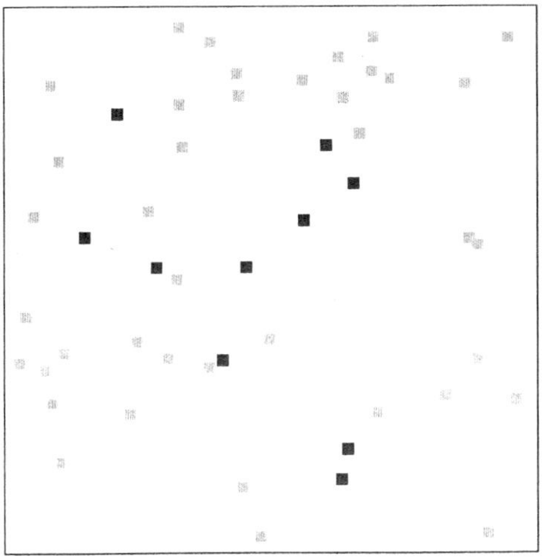

*Figure 6.1.* Distribution of storage locations (black) and transfer points (gray).

sponding number of vehicles (volumes $L_j$ for task $j$) is generated from a uniform distribution with mean $\overline{L}$.

Assuming that the total capacity of the simulated terminal $\hat{K}$ is fully utilized over the entire time horizon, the mean inter-arrival time of storage tasks is $\Lambda = (\overline{L} \cdot \Delta)/\hat{K}$, with $\Delta$ being the intended duration of storage. In the simulation run a storage task $j$ is assigned to period $s_j = t$ determined by rounding its arrival time up to the next largest period number.

**Generating Retrieval Tasks.** After a storage task $j$ is assigned to a period, a storage areas $i$ of sufficient capacity is chosen at random, where $j$ is placed with volume $L_j$. Whenever $j$ cannot be placed anymore because all capacity constraints are violated, a storage areas $i$ is selected at random where $j$ could be placed if the location were empty, i.e. $L_j \leq K_i$.

To place $j$ in $i$, storage capacity is freed by generating retrieval task $k$ in order to remove vehicles from storage area $i$. The number of vehicles to be removed $L_k$ equals the number of its storage task counterpart, i.e. $L_k = L_{V_k}$. Retrieval tasks are contingently generated for location $i$ and assigned to period $t$ until the initiating storage task $j$ can be placed in $i$. Finally, the number of vehicles $L_j$ of $j$ is added to $i$.

To summarize, the intended duration of stay $\Delta$ determines the frequency of generating storage tasks. The capacity limitation of the terminal entails the generation of retrieval tasks at the same rate.

**Altering the Inventory Level.** Since retrieval tasks are generated reactively, the terminal is almost fully utilized and therefore its overall inventory level is close to 1.0. To produce test problems of differing character, we allow a lowering of the overall inventory level with hindsight. Once a retrieval task $k$ has been assigned to period $s_k$, we can safely pre-draw $k$ for a number of periods without risking violation of inventory constraints, i.e. $s_k = s_{V_k} + \lfloor (s_k - s_{V_k}) \cdot \Gamma \rfloor$ with $0 < \Gamma \leq 1$ being the intended inventory level.

Time-windows are generated for tasks $j \in A$ by setting $EST_j = s_j$ and $LFT_j = EST_j + t$, with $t$ uniformly distributed in $[0, (2 \cdot \Delta \cdot \Omega)]$. Thus, the extension of time-windows of tasks are specified in proportion $0 < \Omega \leq 1$ of the indented duration of stay $\Delta$. The existence of time-windows allows the execution of storage tasks later than performed in the simulation run. This gives the opportunity to further decrease the overall inventory level as a matter of optimization.

**Deriving a Problem from the Simulation.** To produce a problem of $T$ periods, the simulation is run for a much larger number of periods in order to prevent initial distortions.

In order to derive a problem instance, all tasks with $EST_j < 100$ are executed as suggested by their $EST_j$. The inventory levels of the storage areas $i \in F : H_i = \mathtt{I}$ at the beginning of period 100 are assigned as initial inventory level $B_i$ for the problem instance. The tasks executed to produce $B_i$ are discarded from being considered in the test problem.

All retrieval tasks $j \in A$ with $EST_j \geq 100$, whose corresponding storage tasks $V_j$ have been discarded because of $EST_{V_j} < 100$, are assigned the storage area chosen for $V_j$ as their prescribed origin $Q_j$. In this way these retrieval tasks completely withdraw the initial inventory level $B_i$ over the course of the simulation.

All tasks with $j \in A : LFT_j > E$, with $E = 100 + T$ are discarded from consideration, merely the destination $Z_j$ of task $j$ is kept to allow the determination of suitable storage area for the corresponding storage task $V_j$ being part of the test problem with $EST_{V_j} \leq E$. Finally all tasks with $EST_j \leq E$ and $LFT_j > E$ are assigned $LFT_j := E$.

For all the tasks $j \in A$ generated, transfer points are chosen at random. Transfer points are determined as origin $Q_j$ for storage tasks and as destinations $Z_j$ for retrieval tasks.

To derive a test problem from the above simulation warrants that a feasible solution exists in principle. Obviously, the information about the solution, i.e. the storage area of tasks chosen as well as their assignments to periods are not passed on to the test problem.

### 6.2.2 Parameterizing Test Problems

In this research we consider problems consisting of more than 1,000 tasks. For these problems we prescribe different inventory levels and we vary time-windows of tasks in order to validate our algorithmic approach.

The terminal is generated with $|\{i \in F : H_i = \mathtt{I}\}| = 10$ storage area and $|\{i \in F : H_i = \mathtt{E}\}| = 30$ transfer points. The capacity $K_i$ for a storage area is drawn from a uniform distribution in $[100, 1900]$. The total capacity of the terminal is summed up to $\hat{K} = \sum_{i \in F : H_i = \mathtt{I}} K_i = 13529$.

All problems consist of $T = 50$ periods with an intended duration of storage of $\Delta = 8.0$. The task volumes are generated from a uniform distribution $[50, 250]$ with mean $\overline{L} = 150$. The mean inter-arrival time $\Lambda$ is determined by $(150 \cdot 8.0)/13529 = 0.088$, such that per period on average 11.3 storage tasks are generated. Since retrieval tasks are produced reactively at the same rate, a problem consists of approximately $22.6 \cdot 50 = 1,130$ tasks.

We vary the overall inventory level $\Gamma \in \{0.7, 0.8, 0.9\}$ to produce a modestly to heavily utilized terminal. The mean extension of time-windows (fraction of $\Delta = 8.0$) is prescribed by $\Omega \in \{0.000, 0.125, 0.250, 0.375, 0.500\}$. For each combination of $\Gamma$ and $\Omega$ we generate problems by varying the seed $\Sigma \in \{1, \ldots, 30\}$. Each of the 450 test problems is uniquely referred to by $(\Gamma, \Omega, \Sigma)$.

## 6.3 Construction Heuristic

### 6.3.1 Greedy Strategy

A low overall inventory level allows choosing among storage area. To support this, storage tasks are assigned to their latest permissible period of processing, whereas retrieval tasks are correspondingly assigned to their earliest permissible period. In order to minimize the overall manpower demand, then the location with the smallest sum of storage and retrieval distance is greedily chosen. We assume this procedure to produce a manpower-efficient solution and refer to it as "greedy strategy" in the following.

From this "greedy strategy" we can expect superior solutions in terms of the of the average manpower demand per period, i.e. $\overline{p} := T^{-1} \sum_t p_t$. By setting the level of regular manpower demand to $P_t := \overline{p}$ for all t considered, the standard deviation $v := T^{-1}\sqrt{\sum_t (P_t - p_t)^2}$ provides a measure for the balance of manpower over the periods considered, cf. the formulation of the objective function in equation (6.1). However, there is no rationale that the "greedy strategy" supports the minimization of

*Table 6.1.* Mean manpower demand $\overline{p}$, deviation of manpower demand $v$, and mean inventory level observed $\overline{l}$ obtained from the construction heuristic parameterized with the "greedy strategy".

| $\Omega$ | $\Gamma = 0.7$ | | | $\Gamma = 0.8$ | | | $\Gamma = 0.9$ | | |
|---|---|---|---|---|---|---|---|---|---|
| | $\overline{p}$ | $v$ | $\overline{l}$ | $\overline{p}$ | $v$ | $\overline{l}$ | $\overline{p}$ | $v$ | $\overline{l}$ |
| 0.000 | 6343.2 | 1209.6 | 0.67 | 6501.8 | 1294.9 | 0.77 | 6756.3 | 1511.5 | 0.87 |
| 0.125 | 6216.5 | 1186.0 | 0.55 | 6324.2 | 1288.2 | 0.65 | 6470.0 | 1348.2 | 0.75 |
| 0.250 | 6137.4 | 1257.9 | 0.46 | 6220.8 | 1276.8 | 0.55 | 6317.1 | 1314.1 | 0.64 |
| 0.375 | 6053.7 | 1293.2 | 0.38 | 6133.6 | 1320.1 | 0.47 | 6201.1 | 1372.2 | 0.56 |
| 0.500 | 5986.8 | 1315.2 | 0.32 | 6054.1 | 1347.2 | 0.41 | 6116.7 | 1400.7 | 0.49 |

$v$, since the opportunity to balance the usage of manpower by altering the period assignment is not used.

Table 6.1 shows the mean values observed over $\Sigma = 30$ test problems of an attribute combination, i.e. the intended inventory level $\Gamma$ and the extension of time-windows $\Omega$, cf. Section 6.2.2. Furthermore, $\overline{l} = (T \cdot \hat{K})^{-1} \sum_{t=1}^{T} \sum_{i \in F: H_i = \mathrm{I}} l_{t,i}$ reports the mean overall inventory level observed. Since the problems are generated by means of a simulation, the observed inventory level $\overline{l}$ is slightly lower than intended by $\Gamma$ even if no time-windows are provided, i.e. $\Omega = 0.000$.

For $\Omega > 0$ we observe a significant decrease of $\overline{l}$, because the number of direct transshipment (without intermediate storage) performed by the "greedy strategy" increases with increasing time-windows. This in turn causes a decrease of the overall inventory level, because the intermediate storage of vehicles is avoided. The manpower demand $\overline{p}$ directly benefits from the greater availability of storage capacity, leading to shorter relocation distances obtained. This observation becomes apparent with increasing load, i.e. $\Gamma = 0.9$ and increasing $\Omega$.

For a higher utilization of the storage capacity the peaks of unbalanced manpower demand become more prominent leading to in increasing deviation of the manpower demand $v$. With respect to an increasing $\Omega$, $v$ first decreases due to the greater availability of storage space, resulting in less prominent peaks of the manpower distribution. While setting $\Omega$ even larger, the "greedy strategy" tends to produce unfavorable decisions for extremely large time-windows, causing a slight increase of $v$ with large $\Omega$.

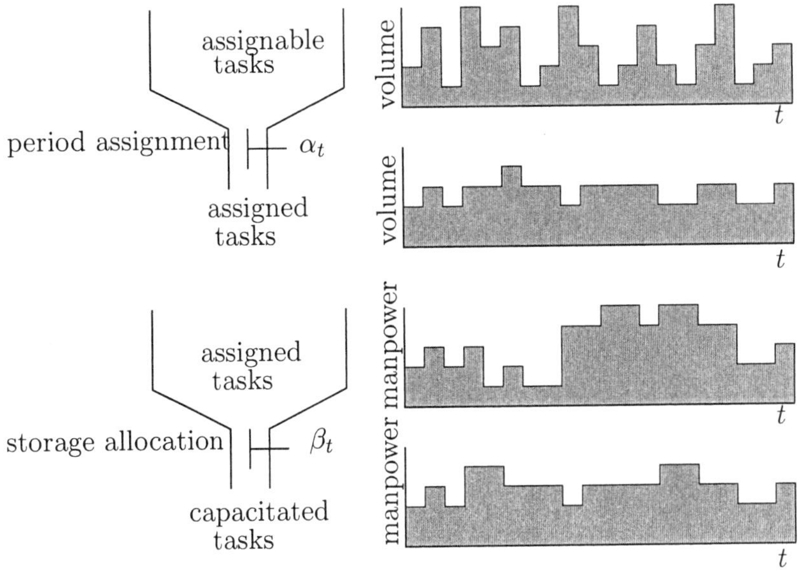


*Figure 6.2.* Scheme of the two-pass construction heuristic.

### 6.3.2 Generalized Procedure

Besides a small overall manpower demand pursued by the "greedy strategy", a balanced manpower utilization over the periods considered is an important goal of the terminal management.

- A balanced load ensures reliable and safe operations and simplifies the integration prioritized tasks which may arrive late. An earlier period of storage and/or a later period of retrieval can be advantageous, even if vehicles unnecessarily occupy storage space by taking this option.

- The portion of the overall relocation distance covered by one task can be a crucial issue for planning. One may accept a reasonable higher manpower consumption — resulting from a long distance for storage — for periods of modest manpower utilization. This decision turns out advantageous if it yields a short retrieval distance for a forthcoming congested period.

In order to allow a variation from the above described "greedy strategy", in the remainder of this section we propose a generalized procedure to be controlled by a set of parameters. The procedure is defined in two passes, illustrated in Fig. 6.2.

An assignment of tasks to periods in accordance to the "greedy strategy" will lead to an unbalanced volume as sketched in the uppermost histogram. In order to balance the volume, we apply a filter on the task-to-period assignment by means of parameter $\alpha_t$. Its appliance can lead to a more balanced distribution of volume as sketched in the second histogram from top. On the upper-left-hand side of Fig.6.2 we depict a possible result of this parameter using the "funnel model" proposed for the load oriented production control developed by Wiendahl (1987) (similar ideas have been proposed in the "optimized production technology" by Goldratt (1997)).

In the next pass we consider the distribution of manpower instead of the volume, compare the y-axis of histograms. The manpower needed is largely determined by the allocation of storage areas. By greedily allocating storage space a balanced distribution of volume may turn into an unbalanced distribution of manpower (depicted by the transition of the second to the third histogram from the top of Fig. 6.2). This time we filter with respect to the manpower demand by means of the parameter $\beta_t$. Finally, we end up with "capacitated tasks" in a hopefully well balanced distribution of manpower as depicted in the lower-most histogram.

**Task-To-Period Assignment.** In a first pass the tasks are assigned to periods. Parameter $\alpha_t$ provides a way to shift task volumes between periods in order to achieve balanced transshipment volumes. Thereby $\alpha_t$ decides upon the number of optionally executable tasks to be processed in period $t$. A large value of $\alpha_t$ causes the execution of almost all assignable tasks, whereas a small value of $\alpha_t$ tends to defer assignable tasks to forthcoming periods. The assignment of tasks to periods is performed for each consecutive period $t = 1 \dots T$ separately, in four steps.

1 Set $\mathcal{S}$ of assignable tasks is built from the backlog of tasks already considered in the last period but not yet assigned ($\mathcal{S} = \emptyset$ for $t = 1$). Then $\mathcal{S}$ is updated by tasks with an earliest processing time $t$, i.e. $\mathcal{S} := \mathcal{S} \cup \{j \in A : EST_j = t\}$.

2 All tasks with $LFT_j = t$ have to be assigned to $t$ and therefore unconditionally enter the set of tasks assigned to period $t$, i.e. $\mathcal{R} := \{\mathcal{S} : LFT_j = t\}$. Finally these tasks are deleted from $\mathcal{S}$, such that $\mathcal{S} := \mathcal{S} \setminus \mathcal{R}$ holds.

3 A fraction of the optionally assignable volume (remaining in $\mathcal{S}$) is specified by $\alpha_t \in [0, 1]$, i.e. $\hat{L} := \alpha_t \sum_{j \in \mathcal{S}} L_j$. In this way the manpower demand of period $t$ is controlled indirectly by the vehicle volume $\hat{L}$.

*Table 6.2.* Example of utility $U_j$ for different $t$ and different type of task.

| period $t$ | 1 | 2 | 3 | 4 | 5 |
|---|---|---|---|---|---|
| storage task $j$ with $LFT_j = 5$ and $L_j = 200$ | -1000 | -800 | -600 | -400 | -200 |
| retrieval task $j$ with $LFT_j = 5$ and $L_j = 200$ | 1000 | 800 | 600 | 400 | 200 |

Optionally processable tasks are determined from $\mathcal{S}$, which are $\emptyset$ for $\alpha_t = 0$ and $\mathcal{S}$ for $\alpha_t = 1$. For $\alpha_t \in (0, 1)$ a subset of $\mathcal{S}$ is selected by means of function $\Phi$ controlled by volume $\hat{L}$ and period $t$, i.e. $\mathcal{R} := \mathcal{R} \cup \Phi^t_{\hat{L}}(\mathcal{S})$.

4 Tasks in $\mathcal{R}$ are assigned to period $t$, i.e. $s_j := t$ for all $j \in \mathcal{R}$. Tasks in $\mathcal{R}$ are deleted from $\mathcal{S}$ such that $\mathcal{S} := \mathcal{S} \setminus \mathcal{R}$ holds and $\mathcal{S}$ merely contains the backlog for period $t + 1$. Finally $\mathcal{R} := \emptyset$.

The specification of function $\Phi$ in step 3 remains as an open issue. $\Phi$ selects tasks in $\mathcal{R} \subseteq \mathcal{S}$ with respect to period $t$ and the maximal volume of transshipment $\hat{L}$. We model the selection problem as a knapsack model, which maximizes its total utility with respect to the constrained knapsack volume $\hat{L}$.

A utility $U_j$ is introduced for all tasks $j \in \mathcal{S}$ which aims at a low overall inventory level of the terminal. Storage tasks are deferred in favor of retrieval tasks. With respect to the volume $L_j$ of tasks, large retrieval tasks are preferred to smaller ones while small storage tasks are preferred to larger tasks. Generally, utility $U_j$ expresses the urgency of executing task $j$ of size $L_j$ with respect to the current period $t$.

Processing retrieval task $j$ in its latest permissible period $LFT_j$ relieves the terminal by $L_j$ units of volume. Pre-drawing $j$ to its earliest permissible period $EST_j$ is credited with $(LFT_j - (EST_j - 1)) \cdot L_j$, because the $L_j$ units of volume freed can be utilized by other tasks for another $(LFT_j - EST_j)$ periods. The above consideration leads to the general term $U_j = (LFT_j - (t - 1)) \cdot L_j$.

In the event of a storage task we have to accept a loss of $L_j$ units of storage space by processing the task in its latest permissible period. By pre-drawing a storage task $j$, its $L_j$ units of storage space are unnecessarily occupied for additional periods. This is taken into account by $U_j := (LFT_j - (t - 1)) \cdot -L_j$ which applies in case of storage. As shown in Tab. 6.2, this selection scheme favors retrieval tasks over storage tasks and therefore aims at a low overall inventory level.

Whenever the time-windows of a storage task $j$ and its corresponding retrieval task $N_j$ overlap, the execution of both tasks in the same period is pursued. In the event that the current period $t$ is still smaller than the earliest permissible period $EST_{N_j}$, $j$ is treated as ordinary storage task as already given above. As time proceeds, $t$ becomes equal or greater than $EST_{N_j}$, i.e. $t$ enters the overlapping interval of $j$ and $N_j$. The greater $t$ becomes, the closer $LFT_j$ gets and consequently the more the execution of $j$ is enforced.

Thus, for the overlapping periods the utility of $j$ is determined by $U_j = ((t+1) - EST_{N_j})/((LFT_j + 1) - EST_{N_j}) * -L_j$. Eventually, with $t = LFT_j$, $j$ is unconditionally executed. Similarly, the execution of a potentially overlapping retrieval task $N_j$ is treated. If the storage $j$ has already been processed, but $N_j$ has not yet been executed, $N_j$ is handled like an ordinary retrieval task described above. Otherwise, the utility of $N_j$ is determined by $U_{N_j} = ((t+1)-EST_{N_j})/((LFT_j+1)-EST_{N_j})*L_j$.

To shift the utility values $U_j$ of tasks $j \in \mathcal{S}$ into $\mathbb{N}$, we add $\min_{i \in \mathcal{S}}\{U_i\}+1$. Now the knapsack problem can be formulated with decision variable $x_j$ such that $\{j \in \mathcal{S} : x_j = 1\}$ constitutes a solution to the problem:

$$\max \sum_{j \in \mathcal{S}} x_j U_j \tag{6.13}$$

$$\sum_{j \in \mathcal{S}} x_j L_j \leq \hat{L} \tag{6.14}$$

$$x_i - x_j \leq 0 \quad \forall i, j \in \mathcal{S} : i \neq j \wedge j = V_i \wedge EST_i \geq LFT_j \tag{6.15}$$

$$x_j \in \{0, 1\} \quad \forall j \in \mathcal{S} \tag{6.16}$$

Eq. (6.13) maximizes the utility while (6.14) depicts the capacity constraint. If time-windows of retrieval task $i$ and storage task $j$ overlap and $j$ has not been scheduled already, (6.15) ensure that in case of selecting retrieval task $j$, also its corresponding storage task $V_j$ is selected. Finally (6.16) express the integer condition of the problem.

This mixed integer problem has been implemented using the software package `lp_solve` version 3.2. To solve a knapsack problem entailed from constructing a solution for one of the test problems defined in Section 4 takes just a fraction of a second only. Since one run of the construction heuristic requires to state $T = 50$ knapsack problems, solving each problem to optimality may become computationally prohibitive when used iteratively as a base heuristic inside an Evolutionary Algorithm, as we will see in Section 6.4. For this reason, in the computational investigation performed the mixed integer formulation of the knapsack problem is replaced by a greedy heuristic task selection scheme with almost no loss of solution quality.

**Allocating Storage Space.** This second pass allocates storage space for the tasks involved. Parameter $\beta_t$ allows shifting the manpower demand between periods by controlling the way storage areas are chosen. Deciding upon the manpower demand of a storage task $j$, the manpower demand of its corresponding retrieval task $N_j$ is determined. If the storage and retrieval of a transshipment are performed in different periods $t$ and $u$, $\beta_t$ and $\beta_u$ determine the relative importance of saving manpower in $t$ with respect to $u$ and vice versa. This way parameter $\beta$ controls whether transports to nearby or remote locations are preferably carried out in a period. The steps to perform are described in the following:

1 The dynamic inventory balance equations (6.11) do not track inventory fluctuations within a period. In order to comply the consecutive space allocation of the construction heuristic with the model proposed in Section 6.1.2, for each period retrieval tasks are processed before storage tasks. Thus, storage space is always freed before being reused.

2 In order to process storage task $j$, considered storage area of sufficient capacity $\mathcal{G} := \{i \in F : H_i = \mathrm{I} \wedge K_i - l_{ti} \geq L_j\}$ are identified. To provide a (contingently feasible) solution in every case, a storage area $d$ with unlimited capacity and extremely large production coefficient with respect to all other locations is provided, $\mathcal{G} := \mathcal{G} \cup \{d\}$. The production coefficient of $d$ is set 10 times larger than the mean of coefficients observed in the terminal. This prevents search from using $d$ unnecessarily, although random solutions will make use of $d$.

3 A storage area is determined by $\min_{i \in \mathcal{G}}\{\beta_{s_j} D_{q_j,i} + \beta_{s_{N_j}} D_{i,z_{N_j}}\}$, $\beta_t \in [0,1]$ . Whenever $\beta_{s_j} = \beta_{s_{N_j}}$ for the period $s_j$ of a storage task $j$ and the period $s_{N_j}$ of its corresponding retrieval task $N_j$, the efforts for storage and retrieval account at the same rate and consequently the location of sufficient capacity with the smallest overall distance available is chosen. In case of a comparably larger $\beta_{s_j}$ the storage effort is favored over the retrieval effort. As a consequence, a nearby location will be chosen at the expense of a higher transportation effort for the corresponding retrieval task (and vice versa).

Figure 6.3 provides an example of the storage allocation scheme. Consider a storage task with origin $a$ to be executed in $t = 1$ and its corresponding retrieval task with destination $d$ to be executed in $t = 2$. In comparison to $t = 1$, manpower capacity is more constrained in

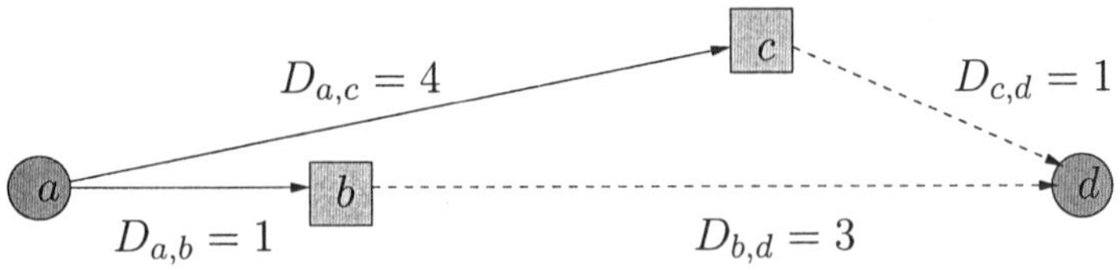


*Figure 6.3.* Example of selecting a location.

$t = 2$. This is expressed by the setting with $\beta_1 = 0.4$ and $\beta_2 = 0.8$ respectively. Two storage areas $b$ or $c$ of sufficient capacity can be chosen for intermediate storage. For the example, due to the setting of $\beta_1$,$\beta_2$ location $c$ is selected with $\min_{i \in \{b,c\}} \{\beta_1 \cdot D_{a,i} + \beta_2 \cdot D_{i,d}\}$. Here, $\min\{(0.4 \cdot 4 + 0.8 \cdot 1), (0.4 \cdot 1 + 0.8 \cdot 3)\} = \min\{1.6 + 0.8, 0.4 + 2.4\} = 2.4$ favors the alternative with the larger sum of productivity coefficients with respect to the higher relevance of manpower productivity in $t = 2$.

4 Finally, inventory is tracked by freeing storage space $l_{t,q_j} := l_{t,q_j} - L_j$ in case of a retrieval task and allocating space $l_{t,i} := l_{t,i} + L_j$ in case of a storage task.

### 6.3.3 Balancing Strategy

In the following we propose a reasonable setting of $\alpha$ and $\beta$ for the generalized procedure, which retains properties of the "greedy strategy" with respect to its efficiency, but shows a much better behavior with respect to the balancing of manpower. Therefore, this strategy is called "balancing strategy" in the following.

Concerning the task-to-period assignment, $\alpha_t, t \in 1, \ldots, T$ has to be set such that tasks are assigned almost evenly distributed over the planning horizon. Parameter $\alpha_t$ applies to the dynamically changing set of assignable tasks. Since the size of this set predominantly depends of the mean extension of the time-windows of the tasks involved, this figure is taken into account by setting $\alpha_t = 1/\Delta$ identically for all periods with $\Delta$ being the intended duration of stay, cf. Section 6.2.

As an example take a problem instance of 10 periods without time window constraints for the tasks involved. Since we aim at scheduling 1/10 of the tasks in every period, we are going to set $\alpha_t = 1/10$ for all $t$ considered. In this way, the task volumes are almost evenly prorated over the planning horizon while the task-to-period assignment procedure takes care of executing retrieval tasks before storage tasks whenever possible.

*Table 6.3.* Improvement in percent obtained by the "balancing strategy" compared to the "greedy strategy" for the mean manpower demand $\overline{p}'$ and the deviation of the mean manpower demand $v'$.

| $\Omega$ | $\Gamma = 0.7$ | | | $\Gamma = 0.8$ | | | $\Gamma = 0.9$ | | |
|---|---|---|---|---|---|---|---|---|---|
| | $\overline{p}'$ | $v'$ | $\overline{l}$ | $\overline{p}'$ | $v'$ | $\overline{l}$ | $\overline{p}'$ | $v'$ | $\overline{l}$ |
| 0.000 | 0.00 | 0.00 | 0.67 | 0.00 | 0.00 | 0.77 | 0.00 | 0.00 | 0.87 |
| 0.125 | -0.12 | 8.82 | 0.57 | -0.16 | 8.35 | 0.67 | -0.56 | 3.59 | 0.77 |
| 0.250 | 0.32 | 9.62 | 0.47 | 0.23 | 9.62 | 0.57 | 0.01 | 7.32 | 0.66 |
| 0.375 | -0.14 | 17.33 | 0.41 | -0.16 | 14.33 | 0.50 | -0.25 | 13.15 | 0.58 |
| 0.500 | 0.23 | 14.18 | 0.35 | 0.20 | 16.65 | 0.44 | 0.10 | 14.66 | 0.52 |

Concerning the allocation of storage space the greedy storage selection scheme is applied by setting $\beta_t = 1.0$ for all periods considered. The regular manpower demand $P_t$ has to be set appropriately. In Figure 6.1, $v$ reports the standard deviation, i.e. the deviation of individual figures against the mean $\overline{p}$ observed. For the "balancing strategy" (and later on in this chapter, for the "adaptive strategy") we proceed in the same vein. Since the "greedy strategy" works effective with respect to $\overline{p}$, we set $P_t = \overline{p}$ as observed for the "greedy strategy". This choice supports our goal to improve the balance of manpower over the periods without increasing the sum of manpower needed at the same time.

Table 6.3 presents the results obtained from applying the "balancing strategy". Instead of $\overline{p}$ and $v$, here the improvements in % over the figures given in Table 6.1 are given as $\overline{p}'$ and $v'$ respectively. In so doing we focus on the advantages of planning over the "greedy strategy". While the sum of the manpower demand $\overline{p}'$ is kept almost constant, the balancing of manpower can be significantly improved with increasing time-windows to approximately 15% for all $\Gamma$ considered. For greater time-windows we observe a saturation of the improvement, particularly for $\Gamma = 0.7$, where the rate of improvement starts to decrease for a large $\Omega$. Interestingly, the utilization of the storage area $\overline{l}$ observed is only slightly larger then observed for the "greedy strategy" in Table 6.1. Obviously, only a small variation of the "greedy strategy" is able to produce a significant reduction of the deviation of manpower over the periods.

In the remainder of this article we investigate whether an individual setting of $\alpha$ and $\beta$ by means of an Evolutionary Algorithm can improve the balancing of manpower even further.

## 6.4 An Evolutionary Algorithm Approach

We briefly introduce Evolutionary Algorithms as a promising candidate for this research, as they have proven successful for many parameter setting problems. Their main advantage is the generic optimization model, which can be applied without further knowledge of the underlying problem to be solved.

### 6.4.1 Evolutionary Algorithms

Evolutionary Algorithms are iterative stochastic search methods based on the principles of natural evolution. They mimic the process of evolution as it was stated by Darwin (1809–1882) in the late 19'th century. The analogy to natural phenomena is best carried out by way of metaphor. Therefore we introduce the basic concept of Evolutionary Algorithms in terms of evolutionary genetics, see Smith (1989).

> "Due to Darwin, individuals with characteristics most favorable for survival and reproduction will not only have more offspring, but they will also pass their characteristics on to those offspring. This phenomenon is known as natural selection."

An individual's characteristics may be advantageously compared to the characteristics of other individuals of the species. These advantages are a relative measure called fitness. The fitness of an individual depends on how its characteristics match the environmental requirements. For the Evolutionary Algorithm, a solution, and in our case a parameter setting, is taken as an individual. Since we assume the same global environment, i.e. the problem to be solved, for all individuals, this population slowly evolves towards individuals of higher fitness by means of natural selection.

The individual's fitness is determined by its acquired characteristics, called its phenotype. The phenotype itself is determined by the individual's genetic prerequisites, called its genotype. Only genotypic information is inherited to offspring. Hence we understand an evolutionary process of a species as a continuous change of genetic material over time. Since Evolutionary Algorithms are inspired by nature, the lingo used in the following is taken from biology.

Genetic Algorithms were developed by Holland (1975), and his associates in the late sixties. Holland referred back to the basic research of Mendel (1822–1884) on genetic inheritance. Therefore he distinguishes between the genotype and the phenotype of an individual. Genetic Algorithms model sexual reproduction by forming offspring from genotypic information of parent individuals.

```
algorithm Genetic Algorithm is
    t := 0
    initialize P(t)
    evaluate individuals in P(t)
    while not terminate do
        t = t + 1
        select P(t) from P(t − 1)
        recombine individuals in P(t)
        evaluate individuals in P(t)
    end while
end algorithm
```

*Figure 6.4.* GA as proposed by Holland.

In Holland's approach a system of continuous variables of a function to be optimized is coded in a binary vector, called a chromosome. A chromosome consists of a finite number of genes with values from the alphabet $\{0, 1\}$. In the Genetic Algorithm lingo we call positions within this vector *loci* and the possible values *alleles*. For fitness evaluation the chromosome is transformed into an argument of the function to be optimized, namely its phenotype. Then, the fitness is determined by means of the objective function.

New chromosomes are generated syntactically by so called genetic operators, which do not use problem specific information. The backbone of genetic search is the crossover operator. It combines the genotypes of two parents in the hopes to produce an even more promising offspring. The logic of the crossover operator assumes that a successful solution can be assembled from highly adapted pieces of different chromosomes. About one half of the genotypic information of two mating individuals is recombined to form an offspring.

In Genetic Algorithms the mutation operator plays a background role. A gene once lost by accident from the gene pool of the population will never appear again. Thus mutation slightly changes chromosomes in order to reintroduce lost genes. Again, mutation works without problem specific knowledge flipping a small number of alleles randomly.

Figure 6.4 gives a brief Genetic Algorithm outline adopted from Holland (1975). Before we are able to run the algorithm, a suitable problem coding has to be found such that solutions of the entire search space can be represented in a chromosome. In a first step we set the generation counter $t$ to zero. Then the initial population $P(0)$ is filled with chromosomes, which consists of uniformly distributed binary values. We evaluate the fitness of all chromosomes in $P(0)$. The evaluation procedure decodes a chromosome into its phenotype and determines its fitness by means of the objective function value. Now we start a loop for a number of generations until some termination criterion is met. A simple termination criterion is a fixed number of generations. A generational reproduction model in encountered.

In each generation the counter $t$ is incremented. A new population $P(t)$ is selected from $P(t-1)$ by some selection operator. Typically proportional selection, also called roulette wheel selection is used. The chance to place individuals in the new population in generation $t$ is proportional to $f_i/\overline{f}_t$ where $f_i$ is the fitness of the $i$'th individual and $\overline{f}_t$ is the average fitness in $P(t)$. The individuals in $P(t)$ are recombined by crossover and mutation. Finally the individuals in $P(t)$ are evaluated in order to obtain fitness values for the selection in the next generation.

A comprehensive introduction to Genetic Algorithms and their properties is given in Reeves (1993), the standard Genetic Algorithm textbook has been written by Goldberg (1989).

Similar ideas have been taken up by two German researchers, Rechenberg and Schwefel in the late 1960s, leading to the notion of Evolutionary Strategies (Schwefel, 1977). They pursued the optimization of engineering systems by means of continuous parameter optimization. Rechenberg and Schwefel altered a single solution to an engineering problem by a random mutation of its set of parameters. Different to Genetic Algorithms, Evolutionary Strategies work on the basis of a real coding of numbers. Thus, mutations have been formulated in terms of small numerical deviations controlled by gradient information.

Two reproductional models have been developed in the context of Evolutionary Strategies (ES). Given the number of parents $\mu$ and the number of offspring $\lambda$ with $\lambda \geq \mu$, we distinguish the $(\mu, \lambda)$ and the $(\mu + \lambda)$ strategy. The population size is fixed to $\mu$ over the generations. In the $(\mu, \lambda)$ strategy the population of the next generation is formed by the $\mu$ best offspring only, whereas in the $(\mu + \lambda)$ strategy the next population is selected from the parents and the offspring of the current generation.

During the 1990s, GA and ES merged, leading to the notion of Evolutionary Algorithms (EA). Depending on the context of optimization, a discrete or real coding and its corresponding mutation and crossover operators are used (Michalewicz, 1996; Bäck, 1996). In this vein crossover operators have been developed for real coded EAs too. The hypercube crossover operator for example generates offspring uniformly on the diagonal joining both parents, i.e. by doing linear combination for each of the continuous variables independently. I.e. an offspring variable $o$ is generated form parent variables $p_1$ and $p_2$ as follows: $o = p_1 + rand() * (p_1 - p_1)$.

Result of the merging towards EA's is a large number of software libraries, which have led to a further standardization of Evolutionary Algorithms (Fink et al., 1999; Woodruff and Voß, 2002; Dreo et al., 2005). The "adaptive strategy" developed in the following is implemented by making use of "Evolvable Objects", a well established C++ class library (Keijzer et al., 2001).

### 6.4.2 Parameter Optimization

The EA proposed in this research adapts the capacity utilization strategy by means of the parameters $\alpha_t$ and $\beta_t$. The construction heuristic proposed in Section 6.3 is integrated as a base heuristic in order to evaluate the strategies evolved.

In order to evolve suitable strategies, a pretty standard EA design is taken from the "Evolvable Objects" open source software library, available from `http://eodev.sourceforge.net`. The `esea` program provided with section 4 of the tutorial of the library in version 0.9.3 is adopted for the adaptation of a storage allocation strategy. This program has been configured as follows:

- A real coded vector has been chosen for representation purpose consisting of $\alpha_t$ and $\beta_t$ of all $t = 1 \dots T$ periods. All vector elements cover the domain $[0, 1]$.
- Mutations as given in are applied at a rate of $p_m = 0.2$ and alter a vector element by $\epsilon = 0.03$, see Bäck et al. (2000).
- A standard hypercube crossover produces superior results and is therefore applied at a rate $p_c = 0.8$, see Booker et al. (2000).
- A population size of 200 individuals has shown to work sufficiently well. 400 offspring are produced and are subject to $(\mu, \lambda)$ replacement (Deb, 2000).
- A generational reproduction model is run for 100 generations, the 40,000 evaluations performed require approximately 15 seconds on a Pentium IV running at 2.6 Ghz.

Because the construction heuristic may produce infeasible solutions due to capacity restrictions, feasibility can neither be warranted by the EA. Since the use of the dummy storage area $d$ is punished by extremely high relocation costs, over the generations infeasible solutions are driven out by means of selection. For the test problems considered, infeasible solutions occur in the very first phase of adaptation only. In the following the best feasible solution obtained for each GA run is reported.

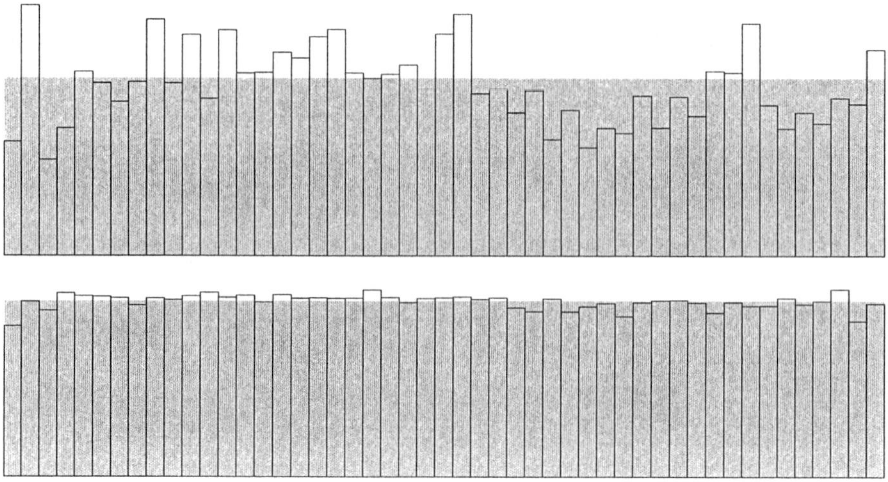

*Figure 6.5.* For a problem of 50 periods solutions are generated by the "greedy strategy" (up) and the "adaptive strategy" (down).

### 6.4.3 Adaptive Strategy

The "adaptive strategy" evolves $\alpha, \beta$ parameter vectors by means of the EA, which are then used to control the generalized procedure. The effect of the "adaptive strategy" against the "greedy strategy" is exemplarily sketched in Figure 6.5 for a single problem instance with $\Gamma = 0.9$ and $\Omega = 0.5$. One clearly observes that manpower is almost perfectly balanced by the "adaptive strategy".

Table 6.4 reports numerical figures obtained for the "adaptive strategy", again given in terms of improvement over results obtained from the "greedy strategy". Generally, drastic improvements with respect to the balancing of manpower rise up to almost 40 %. At the same time, also slight improvements in terms of the sum of the manpower demand $\overline{p}$ can be observed. This finding strikes, since one might pay advances in the distribution of manpower with deterioration with respect to the overall sum of manpower.

Here, major improvements are obtained for test problems without, or with a marginal extension of time-windows. Notice, that the improvements gained for $\Omega = 0.000$ are solely due to the proper setting of $\beta_t$, since in the absence of time-windows tasks are unconditionally assigned without any influence of $\alpha_t$. Thus, merely the choice of locations on a period-oriented basis accounts for up to 28 % of improvement compared to the "greedy strategy".

*Table 6.4.* Improvement in percent obtained by the "adaptive strategy" compared to the "greedy strategy" for the mean manpower demand $\overline{p}'$ and the deviation of the mean manpower demand $v'$.

| $\Omega$ | $\Gamma = 0.7$ | | | $\Gamma = 0.8$ | | | $\Gamma = 0.9$ | | |
|---|---|---|---|---|---|---|---|---|---|
| | $\overline{p}'$ | $v'$ | $\overline{l}$ | $\overline{p}'$ | $v'$ | $\overline{l}$ | $\overline{p}'$ | $v'$ | $\overline{l}$ |
| 0.000 | 0.70 | 28.75 | 0.67 | 0.44 | 28.84 | 0.77 | 0.53 | 28.98 | 0.87 |
| 0.125 | 1.11 | 34.37 | 0.57 | 0.69 | 39.37 | 0.67 | 0.58 | 38.25 | 0.77 |
| 0.250 | 1.17 | 34.18 | 0.51 | 0.85 | 36.68 | 0.61 | 0.37 | 37.03 | 0.71 |
| 0.375 | 0.99 | 33.85 | 0.48 | 0.72 | 36.20 | 0.58 | 0.26 | 36.42 | 0.67 |
| 0.500 | 0.84 | 31.42 | 0.46 | 0.62 | 33.09 | 0.56 | 0.16 | 34.42 | 0.66 |

For $\Omega > 0.000$, the influence of $\alpha_t$ leads to even stronger improvements, although the benefits of controlling the task-assignment seem to be bounded, particularly for larger $\Omega$ values. Interestingly, the mean utilization of storage space $\overline{l}$ is larger than the one observed for the "greedy strategy" and the "balancing strategy", cf. Tables 6.1 and 6.3.

For test problems with large time-windows, the "adaptive strategy" allows storage tasks to be executed early and retrieval tasks to be executed late. This results in a higher utilization of the storage area over time, which may hinder the proper allocation of storage space if used carelessly. Obviously, the GA is able to evolve both controls simultaneously. It yields effective load balancing while retaining efficiency of operations for all variations of the problem data considered.

## 6.5 Summary

We have proposed a multi-period capacitated transshipment problem to improve the efficiency of operations for inter-modal vehicle transshipment. For this problem we have developed an optimization model and we have presented a way to generate challenging problem instances.

We solve these instances by means of an Evolutionary Algorithm which evolves a capacity utilization strategy with respect to the periods considered. The decisions with respect to the individual tasks are performed by an elaborated base heuristic, which is parameterized by a capacity utilization strategy.

We have focused on a period oriented control because the large size of the problem instances prohibits the consideration of tasks as decision entities. We have investigated whether significant improvements over a reasonable deterministic strategy can be gained. The results obtained validate the approach chosen.

This chapter has proposed a solution to the mid-term planning problem of transshipment terminals. This problem domain is insofar interesting as it addresses the derivation of internal work-processes from customer orders. The customer is not concerned about the storage areas chosen to store his goods; his only interest concerning storage is the bridging of time. We have seen that from an operational viewpoint the allocation of storage area is an essential feature for warranting efficient and reliable operations.

Efficiency and reliability can also be improved by exploiting time-windows of customer orders. However, customers will not necessarily offer time-windows for transshipment by themselves. The results obtained for the test problems in this chapter give a strong indication for the importance of time-windows for transshipment planning.

Thus, the management of a transshipment terminal should pay attention to achieve time-windows of operations wherever possible. This issue can be discussed already in the framework-contract closed down with the manufacturer on a strategic/tactical level. Obviously, carrier involved in liner services will not accept a lengthening of discharging or loading because of the internal process optimization issues. However, the retrieval of vehicles to the hinterland transport provides a variety of opportunities for achieving time-windows of operations.

# Chapter 7

# PERSONNEL DEPLOYMENT

**Abstract** Short-term scheduling determines a structure for the employment of personnel. We model such a problem as a multi-mode task-scheduling problem with time windows and precedence constraints. We aim at determining a gang structure supporting reliable as well as efficient operations. For this end we propose a Tabu Search procedure that moves tasks between gangs. In order to determine the manpower demand of a gang we solve the corresponding one-machine scheduling problem by an iterated Schrage heuristic.

## 7.1 The Gang Scheduling Problem

Short term manpower planning integrates personnel into dedicated gangs in order to perform a given set of tasks. This gang structure has to be chosen in a way such that an efficient and reliable processing of tasks is achieved. Generally, a structuring of personnel into gangs will increase the reliability of operations while deteriorating the efficiency of operations because of missing flexibility. We propose an algorithm for determining a gang structure, which ensures the efficient processing of a given set of tasks, c.f. Mattfeld and Branke (2005).

In Chapter 6 tasks have already been assigned to a period and an appropriate storage area has been chosen. Purpose of this chapter is the detailed scheduling of transshipment tasks within single periods. The transshipment tasks assigned to a gang determine its mode of processing, i.e. its minimal level of manpower necessary to perform all tasks assigned to a gang within the planning horizon. Thus, next to the number of gangs incorporated also the tasks assigned to each gang are subject to optimization. We show that the meta-heuristic approach proposed is capable of solving large problem instances reasonably well.

A Tabu Search (TS) procedure moves single tasks between gangs. Whenever a move is performed, its gang of origin and its gang of destination have to be re-scheduled. These sub-problems are modeled as one-machine problems with varying modes of processing. Since such a sub-problem is already NP-hard for a single mode of processing, an efficient base-heuristic is iteratively applied to determine the manpower demand. The selection of such a complex move is controlled by an estimation of the decrease of manpower to be achieved.

Below we discuss related models dealing with task planning under manpower constraints. In Section 7.1.2 we describe the problem at hand in more detail. Then we develop a mathematical model in Section 7.1.3, before we outline a way of generating benchmark problems in Section 7.1.4. In Section 7.2 we describe general properties of Local-Search algorithms before the framework of a Tabu Search procedure is discussed. In Section 7.3 the problem dependent components of the Tabu Search procedure are described in detail, namely the neighborhood definition, the procedure of scheduling a gang, and finally the estimate proposed for selecting a move. We perform a computational investigation for a varying set of problem parameters in Section 7.4. Finally we summarize the potentials of personnel deployment.

### 7.1.1 Literature Review

The problem of assigning manpower to tasks has led to a variety of models and algorithms. In the following we describe related approaches in the order of an increasing level of detail in modeling attributes of manpower and tasks.

**Labor Tour Scheduling.** Dantzig (1954), proposed the labor tour scheduling or shift-scheduling problem. Tasks are modeled as an aggregate in terms of a manpower demand pattern over the time horizon considered. Manpower is assigned by selecting individual shifts, which differ with respect to starting, completion, and relief break times. Labor tour scheduling pursues the minimal number of shifts assigned while covering a prescribed demand pattern. Aykin (2000), extended the original set covering formulation towards the implicit consideration of relief breaks. The resulting decrease of decision variables has opened the way to solve extremely large problem instances today (Mehrotra et al., 2000).

**Workforce Scheduling.** Workforce scheduling also aims at covering a manpower demand by shift assignments, but typically multi-shift scenarios are addressed and it is differentiated between demand profiles and categories of employees. Again, an economical workforce size with respect to the profile specific demand coverage is pursued. In case of the hierarchical workforce scheduling problem categories are arranged hierarchically with respect to employee capabilities. Employees of higher categories may substitute lower categories but not vice versa, see e.g. Narasimhan (1996). In the discontinuous case, employees are considered on an individual basis allowing "days-off" constraints between successive shifts (Brusco and Johns, 1996). Finally, variable demand patterns as already discussed for labor tour scheduling have been taken into account (Billionnet, 1999).

**Machine Scheduling.** Machine scheduling focuses on the detailed modeling of tasks while the availability of resources is depicted in terms of a binary state only. Daniels (1990), has considered a one-machine problem, where tasks can be performed in different modes. Since modes of short duration lead to higher resource consumption, it is aimed at finding resource minimal modes for the tasks involved. For the objective of minimizing the maximum lateness, Daniels (1990), first obtains an optimal task sequence by the earliest due-date priority rule, before he applies a linear program to optimize the resource consumptions of tasks for the given sequence.

**Resource Constrained Project Scheduling.** Resource constrained project scheduling considers the consumption of multiple renewable and non-renewable resources per task (Brucker et al., 1999). The consideration of multiple modes allows a trade-off between a task's processing time and its resource consumption. For prescribed subsets of tasks mode-identity constraints have been introduced in order to allow the assignment of identical personnel to a group of related tasks (Salewski et al., 1997). Finally so-called partially renewable resources have been defined in order to model phases of relief breaks or days-off, see Böttcher et al. (1999).

**Applied Approaches.** Next to the generalized modeling approach of resource constrained project scheduling applied problems have evolved. Airline crew scheduling assigns a crew to a number of subsequent flights, such that all flights are covered and complex safety regulations are met. In doing so the minimization of the number of crews involved is pursued, see e.g. Graves et al. (1993). Driver scheduling refers to a related problem, where shifts have to be selected such that a given number of bus or train routes are covered (Kwan and Wren, 1996).

The audit staff-scheduling problem is the probably most complex one considered so far. Auditors are assigned to engagements each consisting of various subtasks. All subtasks of an engagement have to be performed by the same auditor in prescribed time windows. The duration of subtasks differ depending on the performing auditor. Furthermore, auditors may have phases of non-availability. Finally, auditor dependent setup costs have to be taken into account. Dodin et al. (1998), report on a TS approach for an audit-scheduling problem.

Obviously, an increasing level of detail regarding the consideration of tasks or/and manpower attributes comes along with an increasing number of integer decision variables as well as an increasing number of contingently non-linear constraints. Therefore heuristic approaches dominate algorithms devoted to task planning under manpower constraints.

**Transferred Approaches.** Besides of analogies to models of manpower planning and scheduling, an apparent similarity to the vehicle routing problem with time windows exists (Bramel and Simchi-Levi, 1997, Chapter 7). There, a fleet of vehicles serves a number of customers, even though not every vehicle has to be used. The problem is to assign customers to vehicles, and to generate a route for each of the vehicles, such that a performance criterion is minimal.

Finally, a similar problem is the assignment of non-preemptive computing tasks to groups of processors of a multi-processor system. Tasks can be performed in various modes, the size and number of groups of processors can vary, and finally time-windows for the execution of tasks and precedence relations between computing tasks can exist (Błażewicz and Liu, 1996; Drozdowski, 1996; Feitelson, 1996).

**Container Transport Approaches.** In this book we have chosen a separation of the storage allocation and the personnel scheduling problem. In fact, these problems are interdependent. Particularly for container transports within terminals the duration of transportation tasks is varying. Containers to be relocated have to be assigned to storage areas and transport vehicles have to be assigned to containers for the duration of the container movement. The duration of a container movement depends on the storage area chosen, such that a transport vehicle scheduling model with time varying durations of tasks can be derived. Steenken et al. (1993) describe a real world application, and Chen (1999) presents a simulation model in order to control operations. Böse et al. (2000) suggest a Genetic Algorithm to evlove assignement strategies and Bish et al. (2001) suggest a MIP model for a vehicle scheduling application. For an overview of scheduling equipment and manpower at container terminals see Hartmann (2004).

### 7.1.2 Problem Description

Different to container movements, a vehicle transshipment task can be processed in different modes determined by the number of drivers performing a task, compare Figure 5.5. In order to warrant a safe and reliable relocation of vehicles, drivers are grouped into gangs. The number of gangs as well as their sizes is subject to optimization. However, a gang once established does not change over the course of the planning horizon, i.e. a working shift. Therefore all tasks assigned to one gang are processed in the same mode.

Time windows of tasks and precedence constraints complicate the seamless processing of tasks within a gang. Resulting idle-times equate to a waste of personnel capacity, therefore it is aimed at assigning tasks to gangs such that the personnel capacity of all gangs is seamlessly utilized over the entire planning horizon. This objective can be formulated by minimizing the sum of drivers over all gangs established.

For the problem considered in this research, pair wise precedence relations between tasks are taken into account. This may express that a storage task must be performed before its corresponding retrieval can take place. In addition a precedence relation may express a storage capacity bottleneck between otherwise unrelated tasks. For example, a retrieval task has to be completed clearing a storage area, before another storage task can be started. These real world constraints are concealed by depicting them as pair wise precedence constraints.

As long as the same gang processes two tasks coupled by a precedence constraint, gang scheduling can handle the constraint locally. Whenever a precedence relation exists across gang boundaries, it becomes globally visible, and acts like a dynamically changing time-window constraint for its associated task. In the event that many precedence constraints exist, the seamless utilization of manpower capacity within the various gangs is massively hindered by dynamically changing time-windows. Managing this constraint will be the greatest challenge while searching for a near-optimal solution to the problem.

Although the problem shows many analogies to models discussed above, none of these problem formulations exactly fit the needs of this application. Although already complex models like resource constrained project scheduling with mode-identity constraints or audit-scheduling could be extended towards the requirements of gang scheduling, this additional complexity will prohibit competitive solutions to the problem at hand.

### 7.1.3 Problem Modeling

The model presented in the following refers back to the model given in Section 5.2. Here, we present the model of the multi-mode gang scheduling problem in more general terms.

Let $\mathcal{A}$ be the set of (non-preemptive) tasks involved in a problem. For each task $j \in \mathcal{A}$, a certain volume $V_j$ is to be processed in a time interval specified by its earliest permissible starting time $EST_j$ and its latest permissible finishing time $LFT_j$. The predecessor task of task $j$ of an existing pairwise precedence relation is denoted by $\eta_j$. If no predecessor is defined, $\eta_j = \emptyset$.

Time is modeled by $1, \ldots, T$ discrete time steps, which are treated as periods rather than as points in time. If one task is complete at time $t$, its immediate successor task cannot start before $t + 1$.

The drivers are grouped into a set of $G$ gangs $\mathcal{G} = \{\mathcal{G}_1, \mathcal{G}_2, \ldots, \mathcal{G}_G\}$. Each gang $\mathcal{G}_i$ is assigned a subset of the tasks $\mathcal{A}_i \subseteq \mathcal{A}$ with $\bigcup_{i=1}^{G} \mathcal{A}_i = \mathcal{A}$ and $\bigcap_{i=1}^{G} \mathcal{A}_i = \emptyset$. The number of drivers in gang $\mathcal{G}_i$ is denoted by $p_i$, the number of tasks assigned to gang $\mathcal{G}_i$ is denoted by $h_i = |\mathcal{A}_i|$. At any time step, a gang can only work on a single task.

A solution is described by the number of drivers in each gang, an assignment of tasks to gangs (i.e. a partition of $\mathcal{A}$ into $\mathcal{A}_i$), and a sequence of tasks for each gang.

Let the task on position $k$ in gang $i$'s permutation of tasks be denoted by $\pi_{i,k}$ (i.e. task $\pi_{i,k}$ with $k > 1$ is processed after task $\pi_{i,k-1}$). Starting times of tasks $s_j \in [1, \ldots, T]$ can be derived from such a task sequence by assuming left shifted scheduling at the earliest possible starting time. Similarly, the completion times $c_j \in [1, \ldots, T]$ of tasks are fully determined by the starting times and the manpower demand.

The model can be stated as follows:

$$\min \quad \sum_{i=1}^{G} p_i \tag{7.1}$$

$$s_j \geq EST_j, \quad \forall j \in \mathcal{A} \tag{7.2}$$

$$c_j \leq LFT_j, \quad \forall j \in \mathcal{A} \tag{7.3}$$

$$s_j > c_{\eta_j}, \quad \forall j \in \mathcal{A} : \eta_j \neq \emptyset \tag{7.4}$$

$$s_{\pi_{i,j}} > c_{\pi_{i,j-1}}, \quad \forall i \in \{1 \ldots G\}, j \in \{2 \ldots h_i\} \tag{7.5}$$

$$c_j = s_j + \left\lfloor \frac{V_j}{p_i} \right\rfloor, \quad \forall i \in \{1 \ldots G\}, \forall j \in \mathcal{A}_i \tag{7.6}$$

$$\mathcal{A}_i \subseteq \mathcal{A} \quad \forall i \in \{1 \ldots G\} \tag{7.7}$$

$$\bigcup_{i=1}^{G} \mathcal{A}_i = \mathcal{A} \qquad (7.8)$$

$$\bigcap_{i=1}^{G} \mathcal{A}_i = \emptyset \qquad (7.9)$$

Eq. (7.1) minimizes the sum of drivers $p_i$ over all gangs. Time windows of tasks are taken into account by Eqs. (7.2) and (7.3). Precedence relations among tasks are considered by Eq. (5.4). Eq. (7.5) ensures a feasible order of tasks belonging to the same gang. The completion time of each task is calculated in Eq. (7.6). Finally, Eqs. (7.7) to (7.9) make sure that each task is assigned to exactly one gang.

### 7.1.4 Generating Problem Instances

In order to gauge the quality of our Tabu Search algorithm, challenging benchmark instances are needed. In the following we present a way to produce instances for which optima are known in advance.

The basic idea is to view a schedule as a rectangular plane, where the horizontal extension represents the time dimension and the vertical extension denotes the number of drivers involved. Obviously, an entirely filled rectangle means that all drivers are occupied for the whole planning horizon, i.e. an optimal schedule.

The example in Figure 7.1 shows a schedule with 13 tasks organized in 4 gangs. The planning horizon comprises 25 time units for which 20 drivers are utilized. Time-windows and precedence relations are not yet specified, but can be easily derived from this schedule.

A benchmark instance is generated based on the following input parameters:

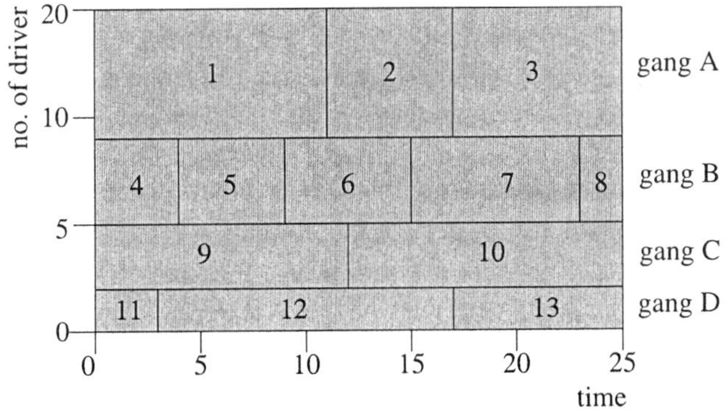


*Figure 7.1.* Exemplary optimal solution to a problem instance.

- the number of time units in the planning horizon $T$
- the total number of tasks $H$
- the number of gangs $G$.
- the minimal and maximal number of drivers in a gang $p_{\min}$ resp. $p_{\max}$
- the percentage of tasks involved in a precedence relation $\gamma \in [0, 1]$
- a parameter $\omega \in [0, 1]$ determining the extension of time windows.

To produce a benchmark instance we proceed as follows:

1 The number of tasks $h_i$ for gang $i$ is determined by prorating the $H$ tasks uniformly upon the $G$ gangs. This can be implemented by first initializing array $K$ of dimension $[0, G]$ and setting $K[0] = 0$ and $K[G] = H$. Then, we assign uniformly distributed random numbers in the range $[1, H-1]$ to $K[1], \ldots, K[G-1]$. Next, we sort K in ascending order, and finally we determine $h_i := K[i] - K[i-1]$.

2 The starting times $s_j$ of task $j \in \mathcal{A}_i$ in gang $i$ are determined by distributing the $h_i$ tasks onto the $T$ time units. We proceed in analogy to the procedure described for Step 1. The completion times $c_j$ of tasks are determined by means of the gang's task chain: $c_{\eta_j} := s_j - 1$.

3 The number of drivers $p_i$ of gang $i$ is drawn uniformly distributed from the range $[p_{\min}, p_{\max}]$. Finally we calculate the task volume by multiplying the task's duration with its manpower demand, i.e. $V_j := ((c_j - s_j) + 1) \cdot p_i$.

4 A task can have a precedence relation with every other non-overlapping task of the schedule. For example, in Figure 7.1 task no. 2 can have a precedence relation with tasks in $\{1, 3, 4, 5, 8, 11, 13\}$. We iteratively select random non-overlapping pairs of tasks not yet involved in a precedence relation and insert a precedence relation until a fraction $\gamma$ of the tasks are involved in a precedence relation or there is no pair of non-overlapping tasks left.

5 Finally time windows are specified in percent $\omega$ of the examined time horizon. In particular, we determine $EST_j := \lceil s_j \cdot \omega \rceil$ and $LFT_j := \lceil c_j + (T - c_j) \cdot \omega \rceil$.

We are able to generate benchmark instances of varying properties with respect to their size, their number of precedence relations, and their time-window extension.

## 7.2 Tabu Search

Tabu Search has been applied to a wide range of combinatorial problems. The technique seems to be generally well suited for highly constrained problems, which allow a neighborhood definition. We are going to show, that Tabu Search is a suitable candidate to solve the gang scheduling problem at hand. Before we describe the algorithmic design in Section 7.3, we sketch the general functioning of Local-Search before we describe the particularities of Tabu Search in some detail. Finally we specify Tabu Search related generic parameters for our gang-scheduling approach.

### 7.2.1 Local-Search Framework

Various Local-Search algorithms have been developed sharing the basic idea of neighborhoods. A neighboring solution is derived from its originator solution by a predefined partial modification, called move (Hertz et al., 1997). A move results in a neighboring solution that differs only slightly from its originator solution. We expect a neighboring solution to produce an objective value of similar quality as its originator solution because both solutions share a majority of characteristics. One can say that a neighboring solution is within the vicinity of its originator. Therefore we concentrate on search within neighborhoods, since the chance to find an improved solution within a neighborhood is much higher than in less correlated areas of the search space.

The success of a Tabu Search procedure heavily depends on the properties of the neighborhood defined. Before we develop a neighborhood definition for the problem at hand, some considerations on desirable features of neighborhoods are addressed (Mattfeld, 1996):

**Correlation** A neighboring solution $s'$ should be highly correlated to its originator $s$. Thus, a neighborhood $\mathcal{N}(s)$ of $s$ should ensure a neighboring solution $s'$ that differs only within a small spread from $s$. This property takes care for a thorough exploration of the search space.

**Feasibility** Perturbations should always lead to feasible solutions. If possible, the search should be restricted to the domain of feasibility in order to avoid expensive repair procedures which in turn would lead to further modifications of $s'$.

**Improvement** A move should have a good chance to obtain an improved $f(s')$ value. In order to achieve this goal additional problem specific knowledge may be incorporated into the neighborhood definition.

**Size** The average size of a set $\mathcal{N}(s)$ should be within useful bounds. A small number of possible moves may halt the search process in early stages (at relatively poor local optima). To the opposite, a large number of moves in $\mathcal{N}$ may be computationally prohibitive if $f$ itself is computationally expensive.

**Connectivity** It should be guaranteed that there is a finite sequence of moves (worsening ones included) leading from an arbitrary schedule to a global optimal one. Otherwise, promising areas of the search space may be excluded from the search. This is known as the connectivity property.

Some of the above considerations may contradict each other. Often these conflicts cannot be solved theoretically. At least some experience with applications is needed in order to develop appropriate neighborhood definitions. Summing up, the features above are desirable properties, which can be used for developing efficient neighborhood definitions.

The simplest deterministic iterative improvement starts from an initial (current) solution, and continually searches the neighborhood of the current solution for a neighboring solution of better quality. Each time a neighboring solution gains an objective value improvement, this neighbor replaces the current solution. The procedure stops if no further improvement can be gained. The described procedure is known as hill climbing in discrete optimization. It can loosely be seen as the counterpart to gradient methods in continuous optimization.

Consider a multi modal objective function. A hill climbing procedure will accept a replacement of the current solution by a neighboring one as long as an improvement can be gained. The final solution is called a local optimum with respect to the neighborhood used. To the contrary, a global optimum is a solution for which the objective value cannot be improved by any other solution of the entire search space. The chance that a local optimum is also a global optimum is very small for most difficult multi modal objective functions. The advantage of having a good chance to improve the objective value within a neighborhood comes along with the drawback of exploring only a small portion of the search space.

In order to avoid the short-come of getting trapped in a local optimum several extensions of the basic hill climbing principle are proposed. They resemble each other in the usage of a more intricate acceptance criterion, which allows a temporary deterioration of the objective value. Such a feature allows the search process to escape from local optima. Examples for such methods are the well-known Simulated-Annealing or the Tabu Search algorithms.

```
algorithm Tabu Search
    build initial solution s and set the best known solution s* = s
    repeat the following steps until a stopping condition is satisfied
        let N(s) be the neighborhood of s w.r.t. the tabu list
        select s' ∈ N(s) for which C(s, s') is minimal
        set s = s' and update the tabu list appropriately
        if f(s) < f(s*), let s* = s endif
    end repeat
    report s* and terminate
end algorithm
```

*Figure 7.2.* A basic Tabu Search algorithm.

The average solution quality obtained by Local-Search strongly depends on the neighborhood definition since the neighborhood definition affects the number of local optima and their distribution in the search space. The search space properties depend on the neighborhood definition applied. Thus, using different neighborhood definitions lead to different appearances of the search space.

If there are only a few local optima present, it's not unlikely that Local-Search will run into a global optimum. On the other hand, if the search runs into a local optimum, the chance to escape is very small even for methods using temporary deterioration. To the contrary, if there are many local optima, hill climbing will perform poor and escape mechanisms are needed. If the local optima are widely spread across the search space, escape mechanisms will more likely fail as if the local optima are closely related.

### 7.2.2 Guidance of Search by Tabu Moves

Glover and Hansen have proposed Tabu Search independently in the 80's. An excellent survey is given in two parts by Glover (1989, 1990). More recent comprehensive treatments can be found in Glover and Laguna (2002); Glover and Kochenberger (2003).

A framework of a Tabu Search algorithm is given in Figure 7.2 with $f()$ being the objective function. Its basic idea of tabu search is to iteratively move from a current solution $s$ to another solution $s' \in \mathcal{N}(s)$ in a neighborhood of the current solution. The neighborhood definition allows "small" changes to a solution, called moves, in order to navigate through the search space. The move to perform is selected upon costs $C(s, s')$ associated with the move from $s$ to $s'$. For the case that the costs are minimized, the deepest descendent, mildest ascendant strategy is applied for selecting moves.

In order to avoid cycling in a local optimum, recently performed moves are kept in a list of currently forbidden moves for a number of iterations. This tabu list is maintained and updated each time a move has been carried out. Unlike Simulated Annealing, an explicit memory of recent moves is kept and evaluated later on for the choice of subsequent moves. Note that keeping moves in memory is not as restrictive as keeping a memory of solutions or parts there of. Not particular points of the search space, but subsets of the path into these points are kept in memory. Glover distinguishes between short- and long-term memory.

**Short-term memory** consists of the last $k$ moves. Typically the short-term memory is implemented as a list, called tabu list. Each time a move is performed, it is stored at the front end of the tabu list. At the same time the $k$'th entry of the list is discarded. If a selected move is part of the tabu list, this move is temporarily forbidden (or tabu in the notion of TS). This mechanism helps to prevent cycles in move sequences after a deterioration of the objective function value has taken place.

**Long-term memory** consists of a data structure keeping track of all moves performed so far. Each time a move is carried out, an annotated counter is increased. The value of the move counter is the basis for a penalty function. The more frequent a move has already been carried out in the past, the more this candidate move is punished. In the long run this penalty avoids searching of areas, which already have been explored. Hence the search is directed into potentially unexplored regions of the search space.

Under certain circumstances the memory may prevent some substantial improvements because it currently forbids a potential good move. Therefore an aspiration criterion is introduced, which temporarily disables the memory function. A useful aspiration criterion is to allow a tabu move if an improvement beyond the best solution found so far can be achieved.

### 7.2.3 Setting of Generic Parameters

Although Tabu Search is generally well suited as a meta-heuristic for the gang scheduling problem at hand, we refrain from specifying an elaborated Tabu Search procedure. In an interactive procedure of iterated plan generation the algorithm used should deliver a reasonable solution within a time span a human planner is willingly to wait for.

Such a time span will be within the range of a minute or less. Thus, moves carried out should contribute with a significant improvement and, the total number of moves to be carried out is limited by definition.

For our purpose we set a move tabu for a variable duration in $[5, \sqrt{N}]$ where $N$ is the number of tasks involved in a problem. Experience shows that this relatively small number suffices to prevent cycling through local optima. Furthermore, a small number of forbidden moves only rarely prevent improving moves from being carried out. Therefore the temporary deterioration by non-improving moves is reasonable small.

For the same reason we do not incorporate an aspiration criterion. The smaller the number of tabu moves is, the more indispensable an aspiration criterion gets. Merely in the very unlikely case that all potential moves are currently set tabu, we perform the most improving tabu move existing.

Since computation time is limited, a stopping criterion of a fixed number of 10,000 moves is used. For a practical application an even smaller number of moves would suffice in order to produce acceptable results. However, a small number of moves carried out prevent a long-term memory from working effectively due to the relatively small number of samples taken. Therefore we confine ourselves to the usage of the recency based short-term memory only.

## 7.3 Algorithmic Design

A solution has to specify three things:

1 the number of gangs $G$ and the number of drivers $p_i$ for each gang $\mathcal{G}_i$

2 the assignment of tasks to gangs, and

3 the sequence of operations in each gang.

In this paper, we are going to use a Tabu Search algorithm to assign tasks to gangs. For every such assignment, inside the Tabu Search heuristic, an appropriate schedule for each gang is derived by applying a simple Schrage scheduler (Carlier, 1982).

The basic idea of Tabu Search is to iteratively move from a current solution $s$ to another solution $s' \in \mathcal{N}(s)$ in the current solution's neighborhood. The neighborhood definition allows "small" changes to a solution, called moves, in order to navigate through the search space. The move to be executed is selected based on costs $C(s, s')$ associated with the move from $s$ to $s'$. For minimization problems, the deepest descent, mildest ascent strategy is applied for selecting moves.

In order to avoid cycling in a local optimum, recently performed moves are kept in a list of currently forbidden ("tabu") moves for a number of iterations. This tabu list is maintained and updated each time a move has been carried out (Glover and Laguna, 1993). We use a variable tabu list length.

In the remaining subsections, the following four issues will be addressed in more detail:

- What is a suitable neighborhood definition?
- How to build an initial solution?
- How to perform a move?
- How to estimate the cost of a move?

### 7.3.1 Neighborhood and Tabu List

The purpose of the Tabu Search framework is the integration and disintegration of gangs by re-assigning tasks. We have chosen the most elementary move possible, namely the re-assignment of a single task from one gang to another, resulting in a neighborhood size of roughly $H \cdot G$ with $H$ being the number of tasks involved in a problem instance. We refrain from engaging more complex neighborhood definitions like the exchange of two tasks between two gangs, because determining the costs $C(s, s')$ for all $s' \in \mathcal{N}(s)$ is computationally prohibitive and, as we will see, the remedy of estimating the costs becomes almost intractable for complex neighborhoods.

As a move attribute we consider a task entering a gang. Consequently, a tabu list entry forbids a task to leave a certain gang for a certain number of iterations. Not every move is permissible. For example, for a certain interval of the planning horizon, whenever there are more tasks to be scheduled than there are time steps available, the move is excluded from $\mathcal{N}(s)$. Also, moves which disintegrate and integrate a gang at the same time are not considered.

### 7.3.2 Building an Initial Solution

The construction of a competitive and feasible solution is not trivial, because the necessary number of gangs is not known in advance. In analogy to the vehicle routing problem described in Section 7.1.2, we build an initial solution by separating tasks into as many gangs as possible. Tasks without precedence relations are placed in a gang of their own, whereas pairs of tasks coupled by a precedence relation share a gang. This way, precedence relations can be handled within the local scope of individual gangs, and it is guaranteed that the initial solution is feasible.

Tasks are placed in a gang, such that the minimal number of drivers required to process the tasks are utilized. In the event that only one task

function *lower_bound*($\mathcal{A}_i$)
    forall $j \in \mathcal{A}_i$ do $p_j = \lceil L_j/((LFT_j - EST_j)+1)\rceil$
    $\hat{p} = \max_{j\in\mathcal{A}_i}\{p_j\}$
    $u = \sum_{t=1}^{T} usable(t, \mathcal{A}_i)$
    $l = \sum_{j\in\mathcal{A}_i} L_j$
    $\hat{p}' = \lceil l/u \rceil$
    return $\max\{\hat{p}, \hat{p}'\}$
end function

*Figure 7.3.* Function *lower_bound*() returns the maximum of two distinct bounds.

is assigned to a gang, or that two tasks share a gang with non-overlapping time windows, for gang $i$ its number of drivers $p_i = \max_{j\in\mathcal{A}_i}\{p_j\}$. The number of drivers required is $p_j = \lceil L_j/((c_j - s_j)+1)\rceil$ where $s_j = EST_j$ and $c_j = LFT_j$.

Whenever two overlapping tasks $j$ and $k$ with predecessor $V_k = j$ form gang $i$, we obtain the number of drivers $p_i$ by treating the overlapping tasks as one task, i.e. $p_i = \lceil (L_j + L_k)/((LFT_k - EST_j)+1)\rceil$. However, the merge of two overlapping tasks may underestimate the number of drivers required due to the integer condition of variables.

Furthermore, we extend the above consideration towards a lower-bound calculation for the number of drivers required for an arbitrary sized set of tasks $\mathcal{A}_i$, c.f. Figure 7.3. First, lower bound $\hat{p}$ refers to the optimistic case that every task can utilize its entire time-window. Then bound $\hat{p}'$ is determined for the case that all tasks overlap by utilizing all possible time steps $t$ of the planning horizon $T$. Function *usable*() returns 1 if time step $t$ can be utilized by at least one task, and 0 otherwise.

The lower-bound will serve as a first optimistic estimate of the number of drivers required in the scheduling procedure subject to Section 7.3.3. Moreover, it is used to produce an initial solution. In this latter case, we contingently increase the number of drivers by 1 until the schedule becomes feasible.

### 7.3.3 Performing a Move

A move is performed in three steps: first, we move the task from one gang to the other. Then, these two gangs are re-scheduled, and possibly their mode (number of drivers) is adapted. Finally, it is checked whether the re-scheduling of the two directly involved gangs can also lead to improvements in other gangs due to changed time-windows. These aspects shall be discussed in the following.

| | |
|---|---|
| function $est(j)$ | function $lft(j)$ |
| $t := EST_j$ | $t := LFT_j$ |
| if $V_j \neq \emptyset$ then | if $W_j \neq \emptyset$ then |
| $t := \max\{t, c_{V_j} + 1\}$ | $t := \min\{t, s_{W_j} - 1\}$ |
| end if | end if |
| return $t$ | return $t$ |
| end function | end function |

*Figure 7.4.* Function $lft()$ returns the latest permissible finishing time of task $j$ w.r.t. to a contingent successor task $W_j$. Function $est()$ returns the earliest permissible starting time of task $j$ w.r.t. to a contingent predecessor task $V_j$.

**Scheduling a Single Gang (in a Given Mode).** Let us first assume that the number of drivers is given. This problem can be seen as a one-machine problem with heads and tails, where the head of a task denotes the non-available interval from $t = 1$ to $EST_j$, and the tail denotes the corresponding interval from $t = LFT_j$ up to the planning horizon $T$. The head of task no. 5 in Figure 7.6 ranges from time unit 1 to 8, whereas its tail comprises only time unit 18. Consequently the time-window of task no. 5 covers time units 9 to 17. In the current mode of processing, two time units are covered.

For our purpose we extend the notion of heads and tails by the consideration of precedence relations of tasks placed in different gangs. Since only one gang is modified at a time, predecessor and successor tasks placed in other gangs may additionally constrain the temporal placement of tasks. The functions $est()$ and $lft()$ in Figure 7.4 restrict the time-window of a task to its currently largest permissible extension. If a predecessor task $V_j$ or a successor task $W_j$ exists, $EST_j$ or $LFT_j$ have to be modified appropriately.

In the event of pursuing the minimization of the makespan the problem has been shown to be NP-hard. For this problem Carlier (1982), proposes a Branch & Bound algorithm, which alters a schedule built by the Schrage-heuristic. We adopt this heuristic by trying to schedule all tasks of $\mathcal{A}_i$ in the planning horizon $1, \ldots, T$ in the given mode $p_i$.

The Schrage-heuristic is sketched in Figure 7.5. Basically, tasks with the smallest permissible finishing time are placed along the time axis. In every iteration, one task is placed starting at time unit $t$. For this purpose all tasks with $est() \leq t$ enter $\mathcal{S}$. From this set the task $j$ with the smallest $lft()$ is selected. If it can be placed at time unit $t$ in mode $p_i$ without violating its time-window, starting and completion time of $j$ are determined, $t$ is updated and finally $j$ is removed from further

```
function schrage(A_i, p_i)
    R := A_i
    t := 0
    while R ≠ ∅ do
        if t < min_{j∈R} est(j) then
            t := min_{j∈R} est(j)
        end if
        S := {j ∈ R : est(j) ≤ t}
        select j ∈ S : lft(j) = min_{k∈S} lft(k)
        s_j := t
        c_j := s_j + ⌈L_j/p_i⌉ − 1
        if c_j > lft(j) then
            return false
        end if
        t := c_j + 1
        R := R\{j}
    end while
    return true
end function
```

*Figure 7.5.* The Schrage-heuristic schedules all task of $\mathcal{A}_i$ in mode $p_i$. Due to the violation of time-windows the function returns "success" or "failure".

consideration. After all tasks have been successfully placed, "true" is returned.

Besides precedence constraints across gang boundaries, precedence relations between tasks to be scheduled in the same gang may exist. The Schrage-heuristic does not take precedence relations into account explicitly. However, regardless of the original problem data we can avoid scheduling successor task $k$ before predecessor task $j = V_k$ by setting $EST_k := \max\{EST_k, EST_j + 1\}$ and $LFT_j := \min\{LFT_j, LFT_k - 1\}$. In order to ensure that $est()$ and $lft()$ handle the precedence of tasks within gangs correctly, we set $s_j := 1$ and $c_j := T$ for all $j \in \mathcal{A}_i$ before the Schrage-heuristic is run.

Figure 7.6 shows a schedule built by the Schrage-heuristic. Initially, tasks no. 1 and 2 can be scheduled at $t = 1$. Task no. 1 is given preference because of its smaller $lft(1) = 3$. In the second iteration only task no. 2 can be placed at $t = 3$. In iteration three, no task is available at $t = 6$ and therefore $t$ is updated to the minimal starting time of the remaining tasks $t = 7$. Task no. 3 dominates no. 4 due to its smaller $lft()$. Then,

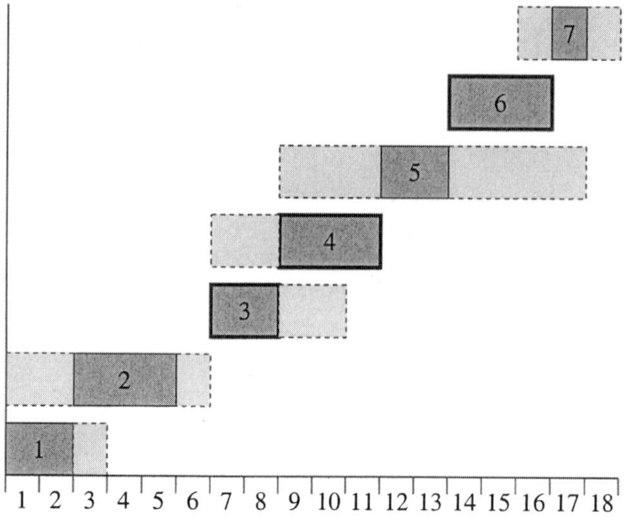


*Figure 7.6.* Example of Schrage-schedule consisting of 7 tasks to be scheduled in 18 time units. Dark gray rectangles represent the time of processing while light gray rectangles depict the time-windows given. Critical tasks are indicated by a black border.

no. 4 is placed at its latest permissible time of placement. Finally the placement of tasks no. 5, 6, and 7 complete the schedule.

**Determining the Manpower-Minimal Mode.** The execution of a move requires solving several gang scheduling problems. In the event that task $j$ is removed from gang $i$, the time units freed may be used to schedule its remaining tasks in a new mode with a smaller number of drivers. To complete the move, the task $j$ removed from gang $i$ has to be inserted into gang $k$. Here, the processing mode may have to change by increasing the number of drivers in order to warrant an integration of task $j$ into gang $k$.

The Schrage-heuristic schedules a gang in a prescribed mode. We aim at scheduling the tasks in $\mathcal{A}_i$ of gang $i$ with the smallest permissible number of drivers $p_i$. To this end we initially set $p_i = lower\_bound(\mathcal{A}_i)$ and contingently increase $p_i$ by one as long as the Schrage-heuristic fails to place all tasks without violating a time-window, c.f. Figure 7.7.

**Propagation of Time-Windows.** Scheduling the tasks of a gang may also entail the re-scheduling of other gangs. In the event that tasks of gang $i$ and $k$ have precedence constraints with tasks of other gangs, a change of $i$'s or $k$'s schedule may impose a change of the time-windows arising from the existence of precedence constraints also for other gangs.

procedure $schedule(\mathcal{A}_i)$
    $p_i := lower\_bound(\mathcal{A}_i)$
    while $schrage(\mathcal{A}_i, p_i) =$ `false` do
        $p_i := p_i + 1$
    end while
end procedure

*Figure 7.7.* Determining the mode of processing which ensures the utilization of the smallest possible number of drivers involved.

Thus, in addition to $i$ and $k$ also other gangs may have to be re-scheduled in order to keep the schedules of the gangs consistent with the existing constraints.

In particular, we appreciate a further extension of a time-window in the event that the completion time $c_j$ of task $j$ impedes an earlier starting time $s_k$ of the successor task $k$ in another gang. Thus, whenever $c_j + 1 = s_k$ holds, and then $c_j$ decreases because the gang of task $j$ is re-scheduled, also the gang of task $k$ is noted for re-scheduling. Similarly, whenever the starting time $s_k$ of task $k$ has impeded a later completion at $c_j$ of the predecessor task $j$ in another gang, $j$ is noted for re-scheduling.

Gangs noted for re-scheduling are held in $\mathcal{P}$ and are determined by function $note()$, see Figure 7.8. Once noted, gangs are re-scheduled in a random order. Since time-windows can be recursively extended, we call this procedure time-window propagation. However, the prerequisites for propagating a time-window are rarely satisfied, such that the number of re-scheduling activities triggered is limited. If, however, gang $i$ is noted

procedure $perform\_move(i, k, j)$
    $\mathcal{A}_i := \mathcal{A}_i \backslash \{j\}$
    $\mathcal{P} := \{i\}$
    $\mathcal{A}_k := \mathcal{A}_k \cup \{j\}$
    $\mathcal{P} := \mathcal{P} \cup \{k\}$
    while $\{l \in \mathcal{P}\} \neq \emptyset$ do
        $schedule(\mathcal{A}_l)$
        $\mathcal{P} := \mathcal{P} \backslash \{l\}$
        $\mathcal{P} := \mathcal{P} \cup \{note(\mathcal{A}_l)\}$
    end while
end procedure

*Figure 7.8.* Performing a move of task $j$ from gang $i$ into gang $k$.

for re-scheduling, there is a reasonable chance to decrease the number of drivers $p_i$.

### 7.3.4 Estimating the Costs of a Move

Since the simulation of a move is computationally burdensome, we estimate the effects concerning the two gangs directly involved in a move and neglect further effects of the propagation of time-windows.

To estimate the costs of a move we determine a contingent saving of drivers $p_i - \hat{p}_i$ due to the removal of a tasks $j$ from gang $i$. Next, we determine the additional effort $\hat{p}_k - p_k$ spent on integrating $j$ into another gang $k$. We calculate the difference of the two figures, i.e. $\hat{p}_i + \hat{p}_k - p_i - p_k$, and select the move with the highest approximated gain (or the lowest approximated losses) for execution.

**Schedule Properties.** In order to estimate $\hat{p}_i$ and $\hat{p}_k$ for gangs $i$ and $k$, we discuss some properties of schedules which will help to derive appropriate estimates. Central notions of our argumentation are the *block* of tasks and the *criticality* of tasks.

DEFINITION 7.1 *A block consists of a sequence of tasks processed without interruption, where the first task starts at its earliest possible starting time and all other tasks start later than their earliest possible starting time.*

Tasks of a block are placed by the Schrage-heuristic independent of all other tasks not belonging to this block. Therefore blocks separate a schedule in distinguishable parts, which can be considered independently. Another interesting property concerning blocks is that slack can occur only at the end of blocks or before the first block.

In Figure 7.6 we identify three blocks. Block 1 consists of tasks no. 1 and 2 and is easily distinguished from block 2 consisting of tasks no. 3, 4, and 5 by the idle-time of time unit 6. Block 3 consisting of tasks no. 6 and 7 can be identified by considering $EST_6 = s_6 = 14$.

DEFINITION 7.2 *A task is said to be critical, if it is involved in a sequence of tasks (called a critical path), which cannot be shifted back or forth in any way without increasing the number of drvivers involved.*

Obviously, all tasks of a critical path belong to the same block, but not every block necessarily contains a critical path. However, if a critical path exists, it starts with the first task of a block. A critical path terminates with the last critical task of its block. Thus, it completes processing at its latest finishing time, although there may exist additional non-critical tasks scheduled later in the same block.

Only one critical path can exist in a block. In the event that a task $j$ scheduled directly before a critical task $k$ causes $k$'s criticality, obviously $j$ itself must be critical. Accordingly, if we classify any task to be critical, all preceding tasks of its block are critical too.

In Figure 7.6 none of the tasks no. 1 and 2 forming the first block are critical, because the entire block could be shifted to the right by one time unit without delaying other tasks. Task no. 3 and 4 form a critical path within the second block. Although task no. 5 is immediately preceded and succeeded by critical tasks, it is not critical by definition, because it does not complete at its latest finishing time. Task no. 6 is again critical without the participation of other tasks placed.

As we will see, the notions of blocks and critical tasks make a valuable contribution to the estimation of a move.

**Estimating the Manpower Release of a Task Removal.** Obviously, every schedule has at least one critical block (a block containing a critical path), which constrains a further decrease of the number of drivers $p_i$. For that reason the only way to obtain a benefit from removing a task from a gang is to break a critical block. If two or more critical blocks exist, and one critical block breaks, at least one other critical block remains unchanged and consequently no benefit can be gained. For instance, in Figure 7.6 the removal of the block consisting of task no. 6 cannot lead to an improvement because tasks no. 3 and 4 still remain critical.

If only one critical block exists, but a non-critical task is removed, again the number of drivers is invariable. As illustrated in Figure 7.9, in all these cases the procedure returns $p_i$. As described below, for the remaining case an improvement of the objective function value due to the reduction of the number of drivers is possible.

**Estimating the Manpower Demand of the Critical Block** Removing a critical task from the only critical block existing, can lead to a decrease of the number of drivers involved. Generally, we prorate the volume of the remaining tasks onto the total time span given. We distinguish the removal of a task 1. within a critical path, 2. at the beginning of a critical path, 3. at the end of a critical path, and finally 4. the removal of a solitary task.

1 In case of the removal of a task inside a path we determine the time-span stretching from the starting time $s_j$ of the first task $j$ of the block to the completion time $c_k$ of the last critical task $k$ of the block. We sum the volumes of all tasks but the one to be removed from the critical path, and divide the sum of volumes through the extension of

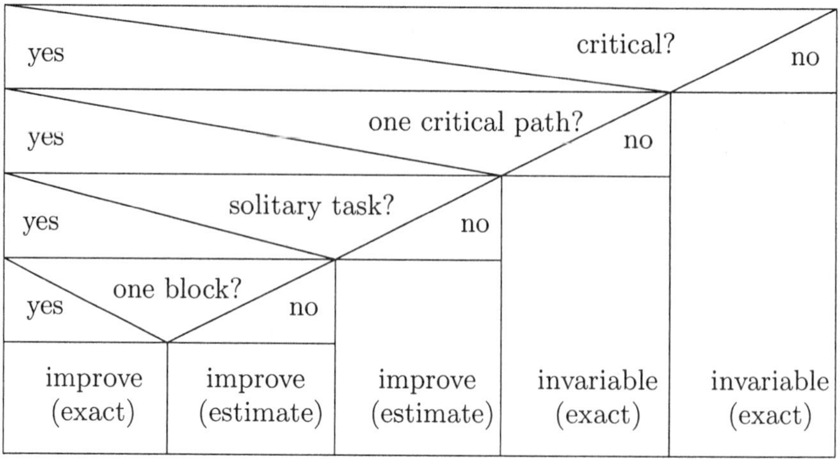


*Figure 7.9.* Scheme of estimating a task removal from a gang.

the time-span. Consider a critical path consisting of tasks 1, 2 and 3. If task 2 is removed, the novel number of drivers is estimated by $(L_1 + L_3)/((c_3 - s_1) + 1)$.

2 The removal of the first task $j$ of a critical path alters the starting condition of the path. Therefore the path can start at the maximum of the earliest possible starting time $EST_k$ of the second task $k$ of the path, and the completion time $c_l + 1$ of the predecessor task $l$ of task $j$ to be removed (if $l$ does not exists, set 1).

3 Similarly, the removal of the last task of a critical path alters the terminating condition of the path. Therefore the path can complete at the minimum of the latest possible completion time $LFT_j$ of the last but one task $j$ of the path, and the starting time $s_k - 1$ of task $k$ succeeding the path (if $k$ doesn't exist, set $T$).

4 In the special case that items 2. and 3. apply at the same time, task $t$ is removed as the last task of a block leading to a manpower requirement of zero.

After having determined the time-span to be considered as well as the volume to prorate onto this time-span, the novel approximated manpower demand for the block is calculated.

**Integrating the Bound Imposed by Non-Critical Blocks** The approximated manpower demand refers to one critical block only. After the removal of a critical task from this block other so far non-critical

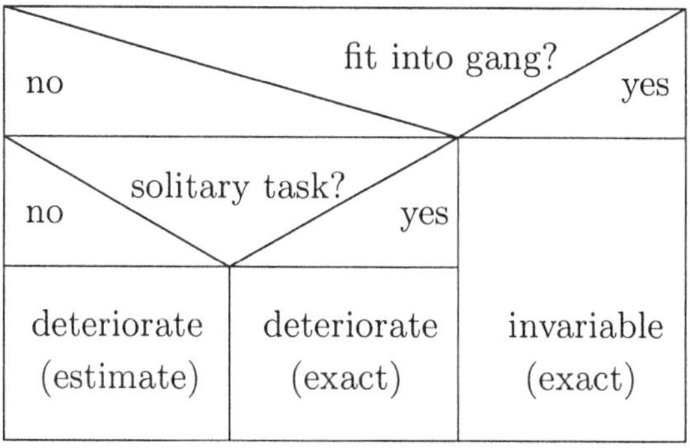


*Figure 7.10.* Scheme of estimating a task insertion into a gang.

blocks may become critical, and for that reason may limit a further decrease of the manpower demand.

Consider a critical block $b_c$ for which the removal of a critical task has been estimated. For blocks in $\mathcal{B} = \{b_1, \ldots, b_{c-1}, b_{c+1}, \ldots, b_n\}$ a lower bound on the number of drivers required is calculated by prorating the sum of volumes of its tasks onto the time-span used by the block plus a contingent idle-time following the block. The number of drivers applicable is then approximated by the drivers $\hat{p}_c$ determined for block $c$ and for the other blocks of $\mathcal{B}$ by $\hat{p}_i = \max\{\hat{p}_c, \max_{k \in \mathcal{B}}\{\hat{p}_k\}\}$.

Together, the procedure accurately identifies the majority of non-improving task removals. If the removal of critical tasks in the only critical block existing may lead to a benefit, this benefit is limited by the manpower capacity required for other blocks. In all these cases a conservative estimate $\hat{p}_i$ is returned by the procedure. Only the improvement of the removal of the last task from a gang can be stated exactly.

**Estimating the Manpower Demand of a Task Insertion.** For estimating the effect on $p_k$ caused by the insertion of a task $t$ into gang $k$, we first try to fit this task into an existing gang schedule. Since we can state a fit exactly, we return the previous number of drivers for this case. Otherwise, we estimate the additional number of drivers $\hat{p}_k$ required in order to produce a feasible schedule. If $t$ integrates a new gang as a solitary task, we even calculate the exact manpower demand $p_k$ of the new gang $k$. See Figure 7.10 for an overview.

**Fitting a Task into an Existing Gang** We start by determining the task $u$ in an existing gang schedule, before the task $t$ is to be inserted. To identify $u$ we scan the tasks of the schedule in the order produced by the Schrage-heuristic and stop at the first task $u$ with $EST_t \leq EST_u$ and $LFT_t < LFT_u$. After having found $u$, we next determine the earliest permissible starting time $s_t$ and the latest permissible completion time $c_t$ with respect to the existing gang structure.

If $t$ has to be appended to the schedule, we easily check whether contingent idle-time $T - c_j$ after the last task $j$ suffices to integrate $t$. For this purpose we set $s_t = \max\{c_j + 1, EST_t\}$ and $c_t = \min\{LFT_t, T\}$.

If $t$ is to be inserted, we are going to verify the available idle-time. Idle-time to the left of $u$ can be available only if $u$ is the first operation of a block. In this case $t$ may start right after the completion time of $u$'s predecessor $v$, i.e. at time step $c_v + 1$. The utilization of the idle-time, however, is limited by $EST_t$, thus $s_t = \max\{c_v + 1, EST_t\}$.

Idle-time on the right can be available only if $u$ is non-critical. In this case $u$ and its non-critical successor tasks can be shifted to the right in order to obtain additional idle-time. The maximal amount of idle-time available can be determined by placing the tasks right-shifted in the opposite order of the task sequence given in the Schrage-schedule. We refer to the resulting starting time of task $u$ as $\overline{s}_u$. Thus, $c_t = \min\{\overline{s}_u - 1, LFT_t\}$.

In the event that $\lceil L_t/((c_t - s_t) + 1) \rceil$ is smaller or equal to the number of drivers currently engaged, task $t$ could be integrated in the schedule without engaging additional drivers. The procedure returns $\hat{p}_i = p_i$ and terminates.

**Extending the Manpower Demand** If the number of drivers does not suffice to integrate task $t$, the additional manpower demand has to be estimated. Since task $t$ for sure becomes critical, the blocks merged by $t$ are considered for prorating $t$'s volume. The approximated increase of the number of drivers $\hat{p}_i$ is returned to the calling procedure.

As a special case of extending the manpower demand we consider the insertion of task $t$ into an empty gang or into a solitary time-window, which does not allow a merge of $t$ with any other task. In this case the exact manpower demand $\hat{p}_i = p_i = \lceil L_t/((LFT_t - EST_t) + 1) \rceil$ is returned by the estimation procedure.

**Tuning of the Estimation Procedure.** Due to the large number of gangs in the initial solution, primarily the disintegration of gangs will prevail. However, in later stages of the search integration and disintegration of gangs will occur counter-balanced.

As a main strength the estimation procedure unerringly recognizes non-deteriorating task insertions. Although important, these cases will

*Table 7.1.* The mean relative error of the local optima found by the local hill-climber.

| | small problems | | | | large problems | | | |
|---|---|---|---|---|---|---|---|---|
| $\gamma/\omega$ | 1.00 | 0.75 | 0.50 | 0.25 | 1.00 | 0.75 | 0.50 | 0.25 |
| 0.00 | 2.5 | 6.3 | 10.4 | 18.0 | 2.1 | 5.9 | 10.4 | 17.3 |
| 0.25 | 2.6 | 6.5 | 10.6 | 16.8 | 2.1 | 6.0 | 10.2 | 16.2 |
| 0.50 | 2.7 | 8.1 | 11.9 | 17.2 | 2.3 | 6.5 | 10.9 | 15.9 |
| 0.75 | 3.3 | 9.9 | 14.6 | 18.6 | 3.1 | 9.1 | 13.6 | 17.1 |
| 1.00 | 5.8 | 13.3 | 17.1 | 20.5 | 6.0 | 12.8 | 16.8 | 18.7 |

rarely occur over the course of the search. In the majority of calls to the procedure, an increasing manpower demand has to be approximated.

Unfortunately, the effect of inserting a task into an existing gang can be approximated rather vague leading to a conservative estimate of the manpower requirement. Only for the integration of a new gang the increase of manpower can be stated exactly. Whenever there is a choice between moving a task into an existing gang and integrating a new gang, the former move will be often preferred by the Tabu Search procedure due to the seemingly better performance.

In addition, the estimation of task removals biases the execution of moves. The saving gained from the disintegration of a gang by removing its last task can be stated exactly. The saving to be gained from removing a task from a gang consisting of several tasks can only be estimated rather vague by a conservative estimate. If there is a choice, the alternative of closing a gang will be often preferred.

Thus, the estimation procedure will unjustifiable favor the disintegration of gangs. In order to subtend this disposition, we multiply the (obviously too small) number of drivers approximated for the insertion of a task into an existing gang with a constant $\delta = 1.1$. In this way the alternative move of integrating a new gang gets a reasonable chance to be executed.

## 7.4 Computational Investigation

To assess the potentials of the Tabu Search procedure, a computational investigation is performed with benchmark problems generated as described in Section 7.1.4.

### 7.4.1 Experimental Setup

We generate problem instances in two sizes, namely with 10 gangs and 100 tasks (small problems) and with 20 gangs and 200 tasks (large problems). For each size, we additionally vary the extension of time windows $\omega$ and the percentage of tasks involved in precedence relations, $\gamma$. Time windows are generated with $\omega \in \{0.25, 0.50, 0.75, 1.00\}$, and the percentage of tasks coupled by a precedence relation are given by $\gamma \in \{0.0, 0.25, 0.50, 0.75, 1.00\}$.

As parameters for the Tabu Search algorithm, we use a variable tabu list length of $[5, \sqrt{H}]$ where $H$ is the number of tasks involved in the problem, and the stopping criterion is fixed at 10,000 iterations.

For every combination of size, time-window extension and number of precedence relations specified, 10 benchmark instances are generated which are solved three times each, because of the non-deterministic nature of the variable sized tabu list length used. Overall, every figure given in Tab. 7.3 and 7.4 represents the mean over 30 samples observed.

### 7.4.2 Hill-Climbing Algorithm

In order to gauge the competitiveness of the Tabu Search procedure, additionally to the Tabu Search procedure a local hill-climber is applied. The algorithm starts from the same initial solution as proposed in Section 3.2 and uses the same neighborhood definition as the Tabu Search procedure, refer to Section 3.1. Different to Tabu Search, the local hill-climber calculates its $C(s, s')$ exactly by simulating all moves in advance. Iteratively, the move yielding the greatest reduction in the number of drivers is performed until a local optimum is reached, i.e. no further improvements can be obtained.

For the local hill-climber, the mean relative error (against the optimal solution) over the 10 benchmark instances of identical parameterization is presented in Table 7.1. For unconstrained problems, the relative error is quite small, with 2.5% and 2.1% for small and large instances respectively. However, with an increasing tightness of the constraints imposed,

*Table 7.2.* Mean number of hill-climbing moves performed.

| | small problems | | | | large problems | | | |
|---|---|---|---|---|---|---|---|---|
| $\gamma/\omega$ | 1.00 | 0.75 | 0.50 | 0.25 | 1.00 | 0.75 | 0.50 | 0.25 |
| 0.00 | 37.4 | 52.5 | 59.2 | 65.9 | 80.3 | 104.8 | 118.5 | 133.9 |
| 0.25 | 31.7 | 45.7 | 51.4 | 59.3 | 70.9 | 91.1 | 103.1 | 120.2 |
| 0.50 | 26.0 | 36.4 | 42.5 | 52.7 | 58.1 | 78.1 | 88.1 | 110.0 |
| 0.75 | 18.8 | 25.2 | 33.9 | 50.2 | 42.7 | 54.2 | 72.9 | 104.8 |
| 1.00 | 3.8 | 11.7 | 28.3 | 47.7 | 10.5 | 27.3 | 55.8 | 99.0 |

*Table 7.3.* The mean relative error of the best solutions found by the Tabu Search procedure.

| | small problems | | | | large problems | | | |
|---|---|---|---|---|---|---|---|---|
| $\gamma/\omega$ | 1.00 | 0.75 | 0.50 | 0.25 | 1.00 | 0.75 | 0.50 | 0.25 |
| 0.00 | 4.59 | 8.86 | 6.87 | 8.67 | 4.71 | 8.76 | 6.89 | 8.77 |
| 0.25 | 4.83 | 8.46 | 7.42 | 8.36 | 4.84 | 8.20 | 8.21 | 9.60 |
| 0.50 | 4.39 | 7.27 | 8.15 | 9.45 | 5.04 | 9.47 | 8.81 | 9.65 |
| 0.75 | 4.32 | 7.71 | 9.93 | 9.81 | 4.95 | 10.32 | 9.97 | 10.70 |
| 1.00 | 3.69 | 9.64 | 9.19 | 10.01 | 4.62 | 9.16 | 9.30 | 10.92 |

the relative error increases drastically to approximately 20% for $\gamma = 1.00$ and $\omega = 0.25$.

Obviously, the search space becomes more rugged, and local optima become more frequent. This goes along with a decreasing performance of the hill-climber. However, as shown in Table 7.2, the number of hill-climbing moves performed does not directly reflect the ruggedness of the space to be searched. As one may expect, the hill-climbing paths get shorter with an increasing number of precedence constraints imposed ($\gamma$). The reasonable relative error of $\approx 6$ % for $\gamma = 1.0$ and $\omega = 1.0$ is obtained by a mere 3.8 moves on average for small problems and 10.5 for large problems respectively.

By narrowing the time-windows from the entire planning horizon ($\omega = 1.0$) towards 1/4th of the horizon ($\omega = 0.25$), the number of moves performed on a downhill walk increases significantly. Apparently, tight time-windows introduce a locality to search such that only tiny improvements per move can be obtained. Although with $\omega = 0.25$ more than 100 moves are performed for large problems, the relative error obtained increases with an increasing tightness of time-windows.

Despite performing an increasing number of moves, an increasing relative error is observed. Thus, a further improvement in searching a rugged search space requires the temporary deterioration of the objective function value. Although the Tabu Search procedure provides this feature, the large number of iterations needed by the Tabu Search procedure requires a considerably faster estimation of move costs. This can be obtained by the estimation procedure developed, although at the expense of inaccurate measures.

*Table 7.4.* Number of gangs recorded with the best solution observed.

| | small problems | | | | large problems | | | |
|---|---|---|---|---|---|---|---|---|
| $\gamma/\omega$ | 1.00 | 0.75 | 0.50 | 0.25 | 1.00 | 0.75 | 0.50 | 0.25 |
| 0.00 | 70.2 | 30.5 | 16.2 | 11.2 | 138.5 | 61.3 | 29.3 | 20.5 |
| 0.25 | 64.6 | 29.1 | 14.1 | 11.0 | 125.5 | 51.3 | 23.9 | 19.4 |
| 0.50 | 57.1 | 20.5 | 14.6 | 11.3 | 115.3 | 54.4 | 27.7 | 19.0 |
| 0.75 | 49.8 | 18.7 | 14.9 | 10.8 | 100.7 | 56.2 | 25.8 | 18.3 |
| 1.00 | 39.5 | 28.8 | 13.3 | 10.5 | 82.9 | 36.1 | 21.4 | 18.4 |

### 7.4.3 Tabu Search Algorithm

Tab. 7.3 presents the mean relative error observed for the Tabu Search procedure. For a maximal time-window extension $\omega = 1.00$ the relative error comprises $\approx 4\%$ regardless of the number of precedence relations specified. This is twice the relative error observed for the hill-climber, and pinpoints at the shortcoming of the estimation procedure. Obviously, the estimation delivers poor approximation of the changes in the number of drivers imposed by a move in the case of loose constraints.

Narrowing the time windows increases the relative error of the Tabu Search procedure only slightly up to $\approx 10\%$ for $\omega = 0.25$. This figure is approximately half of the relative error observed for the hill-climber, which convincingly demonstrates the advantage of Tabu Search for problem instances with tight constraints.

The tighter the constraints are, the better the costs of a move are approximated by our estimation procedure, because the time span onto processing times are prorated decreases with an increasing tightness of constraints. Therefore, the estimation procedure is able to guide the search more accurately. The algorithm proposed seems pleasantly robust against the existence of precedence relations and an increasing problem size.

Tab. 7.4 shows the mean number of gangs recorded with the best solution observed. This figure demonstrates how well the gang structure of the optimal solution has been reproduced by the Tabu Search procedure. For large time-windows ($\omega = 1.00$) an enormous number of gangs have been integrated. Since only the integer condition on the number of drivers restricts the algorithm from finding the optimal solution, there is no need to reduce the number of gangs. However, merely the existence of precedence relations (which has only a small influence on the solution quality) cuts the number of gangs in half.

For higher constrained problems with $\gamma \geq 0.5$ and $\omega \leq 0.5$ the gang structure of the optimal solution is successfully approximated. Here, for small problem instances $\approx 10$ gangs are integrated, whereas the 200 tasks of large problem instances are essentially distributed among $\approx$ 20 gangs. For highly constrained problems an effective gang structure is a prerequisite for obtaining high quality solutions. Obviously, this structure is identified by the algorithmic approach proposed.

## 7.5 Summary

In this chapter we have addressed the problem of finding a suitable gang structure for a task scheduling problem such that the number of drivers required becomes minimal. For this problem we have presented a model and we have proposed a way to generate benchmark instances of varying properties. We have developed an efficient Tabu Search procedure in order to solve such gang scheduling problems.

Particular attention has been spent on the design of the Schrage-scheduler acting as a base-heuristic in the Tabu Search framework. Although the move of a task from one gang to another modifies just these two gangs directly, the other gangs are indirectly affected by the change of time window constraints and have to be rescheduled as well.

Since the move neighborhood is large and the calculation of the cost of benefit is computational expensive, we have proposed a cost estimation procedure, which approximates the outcome of a move before it is actually performed. Although an estimate must be imperfect in the face of the move's complexity, experience has confirmed the applicability of the estimate developed. For a wide range of benchmark instances a promising solution quality has been achieved.

We have shown throughout Chapter 6 and 7 that an efficient algorithmic solution to the sub-problems derived in Chapter 5 is possible at reasonable computation costs. The heuristics defined are fast enough to be applied in an interactive framework controlled by a human planner. In the following we consider the integration of the algorithmic solution into an existing IT-infrastructure based on an enterprise resource planning solution.

# Chapter 8

# IT-INTEGRATION OF PLANNING

**Abstract** In this chapter we discuss the integration of operations planning into an existing IT-infrastructure of a transshipment terminal. By example of the vehicle transshipment terminal Bremerhaven we outline general concepts of integrating planning issues as a re-engineering activity. We start with a description of commonly used IT-functions supporting the execution of terminal operations, before we sketch the interfaces used in order to synchronize the information system level and the execution level.

We then focus on software modules for the user interface to planning. Finally, we discuss details of the integration of planning by viewing the planning activity as a business process. We handle this issue on the level of the requirement definition, the design specification and the implementation description. We conclude this chapter with a discussion of the impact of planning for the vehicle transshipment terminal of Bremerhaven.

## 8.1 Support for Terminal Operations

The need for an integrated IT system to support terminal operations is known since long (Leeper, 1988; Giannopoulos, 2004). Its benefits with regard to an increased terminal performance has been proven, for instance by Wan et al. (1992) for the port of Singapore. Three main functions have to be distinguished: First, the data exchange by means of EDI (Garstone, 1995), second, the tracking of detailed operations by means of an ERP (Kia et al., 2000), and finally, the planning and scheduling of terminal operations as proposed in this book.

So far, we have discussed prospects of operations planning for a transshipment terminal, we have derived suitable optimization models and we have developed efficient heuristics to solve the arising problems. Thus, we are ready to build a software module in the form of a callable optimization library. However, without a proper integration of the terminal operations planning (TOP) into the existing IT environment, we cannot expect any positive impact on the efficiency and reliability of operations.

Clearly, the TOP module has to be supplied with appropriate input data. This sounds trivial, but usually information is hardly available at an aggregate level that is appropriate for optimization. Customer delivered transshipment orders have to be aggregated into (internal) tasks. Moreover, data with regard to the port infrastructure like storage capacities, travel ways and distances have to be provided.

Once a implementation scenario is generated by the TOP module, this solution has to be evaluated an analyzed. Probably multiple optimization runs with modified input data are needed until a solution fully satisfies a human planner. Since the analysis of a solution is by no means trivial, a significant effort has to be spent on the visualization of implementation scenarios. Finally, one accepted planning scenario is implemented by means of a disaggregation of the planning objects into meaningful work-processes for terminal operations.

The aggregation of customer orders into meaningful objects for planning as well as their future disaggregation are typically well supported by enterprise resource planning (ERP) systems. However, the "planning functionality", i.e. the anticipation of decision alternatives, is typically less distinctive (Shobrys and White, 2002). Purpose of the following consideration is the integration of planning into an existing ERP framework by example of the transshipment of finished vehicles at the Bremerhaven terminal, c.f. Mattfeld (2005).

## 8.1.1 Functionality of ERP Systems

Primary goal of an ERP system is the mapping of transshipment processes to the information system in order to provide control for the execution of logistic processes (Rizzi and Zamboni, 1999). For vehicle transshipment this mapping starts with the definition of internal planning tasks from customer orders. The execution of tasks is then controlled by tracing the state of vehicle transshipments. This also comprises the tracking of the storing position of vehicles. After a vehicle has been retrieved from the terminal, its transshipment is accounted by issuing an invoice to the customer. Finally, transshipment operations are controlled with hindsight by means of debit/actual comparisons.

For the Bremerhaven terminal, the above ERP functionality is implemented in the "car individual network"(CARIN) software , developed by BLG data services, a subsidiary of BLG Logistics AG. The name CARIN implies that the system covers the logistics network on the basis of *individual* vehicles. This functionality of CARIN is emphasized by Drewry

(1999): "BLG was amongst the first terminal operators to set up an on-line status tracking system for car exports. This was designed initially for BMW and Mercedes exports to England, the Far East and USA and has since been expanded to include all vehicle movements".

In order to offer more complex transshipment arrangements to the customer, also the tracking and tracing of vehicles beyond the terminal area is supported (Polewa et al., 1997). The CARIN system "is designed to act as a link from the manufacturer's office overseas via Bremerhaven, via the various port status indicators and pre-delivery inspection facilities, direct to the dealer. The customers can thus ascertain the exact position of their cars at any time with BLG providing the communications interface" (Drewry, 1999).

CARIN has recently been extended by means of a web-based interface. Kuhr (2000), states, that CARIN integrates the entire chain of transport from the manufacturing plant in Japan to the import forwarders and all the way to the dealer in the destination country. Although possible from a technical point of view, the organizational obstacles imposed by the large number of firms involved in the logistics network dominate. Thus, a complete trace seems possible only for long-termed arrangements with only a few partners involved in the logistics chain (Nijkamp et al., 1996).

Bubois and Gadde (1997) notice, that in the transport industry freight forwarders and transportation companies have been able to strengthen their positions by establishing efficient large-scale transportation systems as well as local timetable based delivery services. The extension of the ERP system of the BLG can be seen as a contribution in this fruitful direction.

### 8.1.2 Control of Terminal Operations

The tracing of vehicle along the logistics import chain requires interfaces to many external and internal sub-systems. Figure 8.1 shows control interactions between the execution level and the information system. First, an avis of vehicles is received from the manufacturer by a notice of dispatch via electronic data interchange (EDI). Before the respective vehicles are discharged at Bremerhaven, usually also "manifesto" is received from the carrier, describing type and physical conditions of the vehicles to be unload from the calling car-carrier.

Since the responsibility is incurred from the carrier to the BLG at the point of discharge, vehicles are checked for damages etc. when entering the terminal. A bar-code badge is placed inside the front window of a vehicle, which is then associated with the vehicle's identification number (VIN) and other data already received from the manufacturer. In this way the vehicle becomes existent for the ERP system CARIN. After

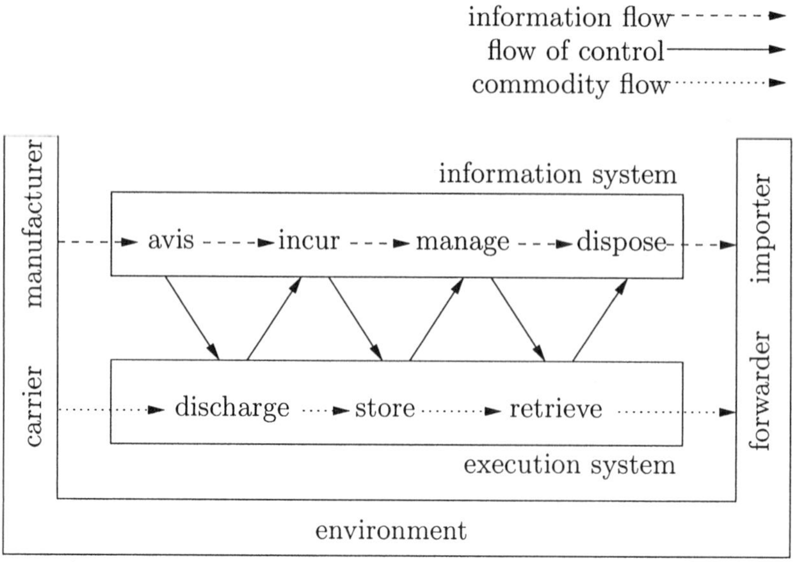


*Figure 8.1.* Interaction of execution and planning activities.

the vehicle has been stored into a storage area, its physical position is transferred to the ERP system for further management by scanning its bar-code badge.

After the release order for a vehicle has been received from the importer or forwarder, the vehicle is retrieved from the storage area and loaded onto rail, truck or feeder ship. Since the responsibility for the vehicle is disposed, the vehicle is once more checked for damages, this time on the behalf of the forwarder. Finally its departure is noted in the ERP system. For controlling purposes, the productivity of transshipment operations is recorded with hindsight.

## 8.2 Software Support for Planning

Although substantial effort has been spent on the design of the ERP system, its use for planning purposes is limited (Galliers, 1994). Data structures suitable for "what-if" considerations are missing, such that planning purposes can hardly be covered. This deficiency is mainly caused by the detailed data structures provided with respect to process control and execution issues.

Furthermore, CARIN covers the terminal infrastructure as the most important resource only rudimentary. Data access paths acquire storing positions via individual vehicle data sets, and not vice versa. In this way the physical position of a vehicle in the terminal can be easily tracked,

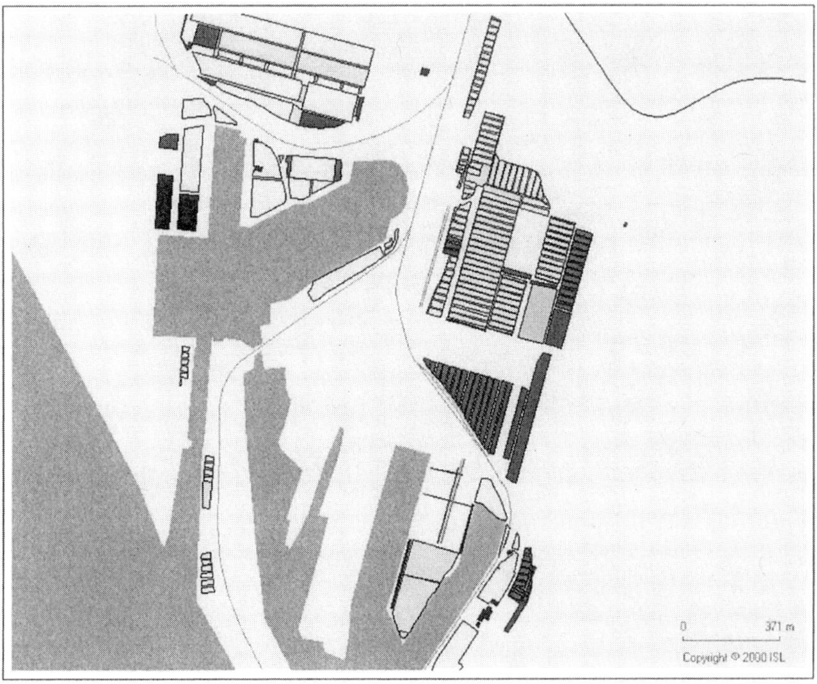


*Figure 8.2.* Overview of vehicle transshipment terminal as it appears in the terminal-viewer. The middle-gray area in the southeast shows the seaside at the Weser River with water connections to the port area. Two large compounds can be identified in the north (primarily export) and in the east of the port (primarily import). The ruler in the southeast of the plot determines a length of 371 meter.

but to determine the number of vehicles stored at a certain area, as demanded by capacitated planning, will be a tedious task.

## 8.2.1 Terminal Viewer Module

In order to integrate the infrastructure into planning, the Institute of Shipping Logistics (ISL), Bremerhaven, has developed a graphical user interface for this purpose. The "terminal viewer" is able to display geo-data regarding travel ways, quaysides, rail ramps, transfer points and storage areas in several scales ranging from a port overview to a detailed mapping of one storage area.

Since an enormous effort is spent for this important prerequisite of resource planning, the activities related to the collection and processing of data are briefly sketched in the following. First of all, a digital map of sufficient resolution is required, which may be trivially purchased for most public properties nowadays, but which has to be individually produced for the majority of private properties (Jiang et al., 2003).

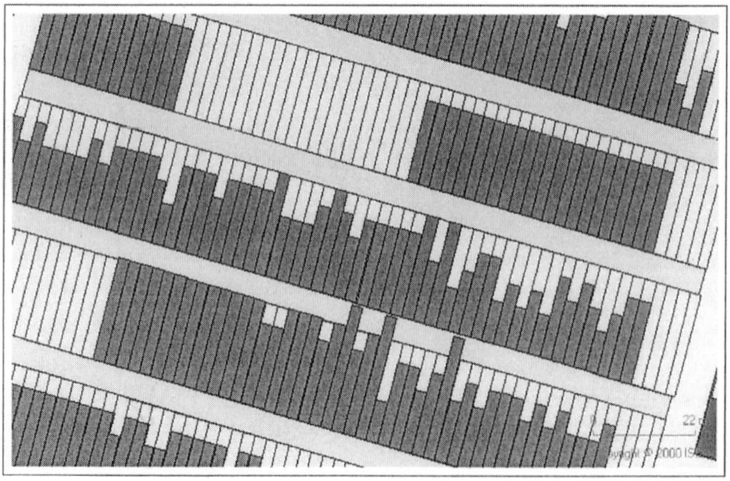

*Figure 8.3.* Detailed view on the capacity utilization of one single storage area as it appears in the terminal-viewer. One clearly recognizes the distinct blocks each consisting of a large number of parallel rows. Rows are appear empty, semi-full, full, or even over-loaded with respect to their prescribed capacity. The ruler in the southeast of the plot shows a distance of 22 meters in this detailed scaling.

A differential global positioning system (DGPS) is used to determine geo-coordinates of the transshipment terminal. DGPS receives satellite signals like a commonly used GPS, but corrects these signals by means of an additional transmitter located at a nearby radio tower. In Bremerhaven the DGPS transmitter of the neighboring container terminal is engaged to collect geo-coordinates deviating from their true location by less than one centimeter. Nevertheless, determining more than 6,000 geo-coordinates is truly a matter of diligence.

In a next step, the coordinates are assigned to objects in order to model spatial borders and separations of objects. Then, the capacity of storage area is estimated by means of the physical extension of areas. However, as can be taken form Figure 8.2, storage areas are for historic reasons by no means rectangular. Thus, again a manual examination of storage capacities is unavoidable.

Storage locations are integrated into a four level hierarchy, starting with storage rows of four to ten vehicles capacity at the lowest level, c.f. Figure 8.3. Rows are integrated into blocks, which are again integrated into areas. Finally areas are integrated into terminal regions at the top level. For planning purposes, the hierarchy level storage space provides a reasonable number of objects of sufficient capacity (approximately 1,000 vehicles per area), which are distinguishable with respect to their assigned productivity.

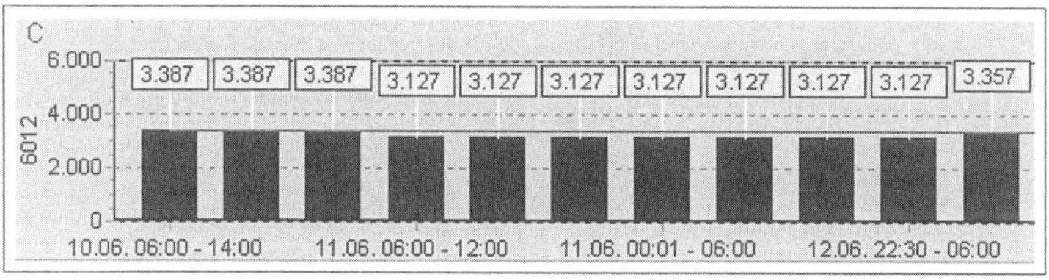


*Figure 8.4.* Utilization of a storage area "C" with a capacity of 6,000 vehicles over eleven shifts of three days.

Distances with respect to travel-ways are calculated in a further step. Therefore the medians of storage areas are determined and, based on this data, the distance between each two of the 80 storage areas and the 60 transfer points considered are determined. This distance metric serves as source for the derivation of productivity measures for optimization.

A database link to the ERP system opens the opportunity to display the capacity utilization of areas as shown in Figure 8.3. Besides a visualization of the actual capacity utilization, targeted debit utilization produced by automated planning can be displayed too. In this way, the terminal viewer can be used as a valuable tool for analyzing planning scenarios.

### 8.2.2 Terminal Information System

A simulated view at the planned state of the terminal will hardly suffice to assess a planning scenario. Therefore a main challenge is the processing of various aspects of a planning scenario with respect to human decision support. A graphical user interface visualizes a planning scenario, so that it can be analyzed and contingently modified by a human planner.

The ISL has also developed such a software system, called auto terminal information system (TIS). TIS receives a set of partially specified tasks from the ERP system, prepares these tasks for optimization, and finally receives a planning scenario, i.e. a solution consisting of fully specified tasks, from the optimization procedure. Dependent on manual modifications of the task data, the optimization cycle may be run for multiple times.

In order to assess a planning scenario, different points of view to the scenario are provided. Next to the terminal viewer, the capacity utilization of single storage areas can be displayed over time. Figure 8.4 gives an example of the time-oriented resource viewer. The utilization of manpower can be displayed in a similar fashion. However, viewing

| Lager | | Status | |
|---|---|---|---|
| Lager | Kapazität | Belegt | Plan |
| BLG Terminal | 16.963 | 8.550 | 10.470 |
| C | 6.012 | 3.387 | 3.227 |
| FLA | 2.413 | 887 | 987 |
| LAG | 2.931 | 2.456 | 2.196 |
| PAR | 81 | 19 | 19 |
| WAS | 587 | 25 | 25 |
| NH | 0 | 0 | 0 |
| AK1 | 0 | 0 | 0 |
| WES | 10.951 | 5.163 | 7.243 |
| A | 490 | 51 | 51 |
| B | 5.424 | 2.858 | 4.938 |
| C | 1.107 | 0 | 0 |

*Figure 8.5.* Hierarchy of storage areas with capacity and storage utilization.

the progression of resource utilization is limited to important figures of central meaning.

For other resources an overview in terms of starting and finishing conditions may suffice. Figure 8.5 shows an example of the development of the utilization of the storage capacity for several locations. The storage areas are structured according to the hierarchy levels of regions and areas. For these objects the capacity and the space allocation are shown at the begin ("Belegt") and end ("Plan") of the planning horizon. The main advantage of this way of presentation is the user-defined level of detail for the various objects depicted.

The display of tasks can be selected with respect to their type, time of processing, and storage or retrieval location, c.f. Figure 8.6. This feature allows the grouping of tasks in order to solve planning conflicts: In Chapter 6 a dummy storage area of infinite capacity is introduced to warrant that the algorithm produces a feasible solution in every case. Obviously, such a solution is not necessarily feasible with regard to practice. Whenever vehicles are place at the dummy storage area, manual intervention is needed. The task viewer provides a selection of mistakenly assigned tasks.

The task editor is provided in order to allow manual modifications of planning scenarios. There are two issues related to the manual modification of solutions. First, changes made in the task editor have to be propagated to terminal, resource, capacity and task viewers. This requires the existence of a central presentation management that controls the different viewers. Whenever data change, it is checked for out-of-date contents in currently open viewers. If out-of-date information is detected, the respective viewer is notified to reload its data content.

A more intricate issue of manual interaction is related to the temporary violation of model constraints. A manual modification may require a chain of modifications, which temporarily drives the planning scenario into an infeasible state with respect to model constraints. Thus, the task editor functionality can merely avoid the violation of constraints with respect to static input data. Significant effort is spent to guide a human planner back to feasible planning scenarios in case of manual modifications.

Together, TIS consists a collection of interacting graphical user interfaces, i.e. the terminal viewer, the resource viewer, the task viewer and finally the task editor. These parts of the user interface are to be configured individually in order to efficiently analyze and modify planning scenarios.

## 8.3 Interplay of Execution and Planning

So far, requisitions of an ERP system suited for the management of a transshipment terminal have been described. Furthermore, light has been shed on the functionality of TIS, suited for the analysis and control of the TOP module. Subject of this section is the requirements planning, the design specification and the implementation description of the integration of automated planning support. Thereby, we view the planning activity as a business process, starting from the dispatch of a customer order and ending with the activities scheduled for execution (Ng et al., 1999). Although many different tools exist to support the development and documentation of business processes, we confine ourselves to models with regard to the ARIS toolset (Scheer, 1999).

### 8.3.1 Requirements Definition

The requirements definition considers the functions and interactions of a business process with respect to different organizational units. In

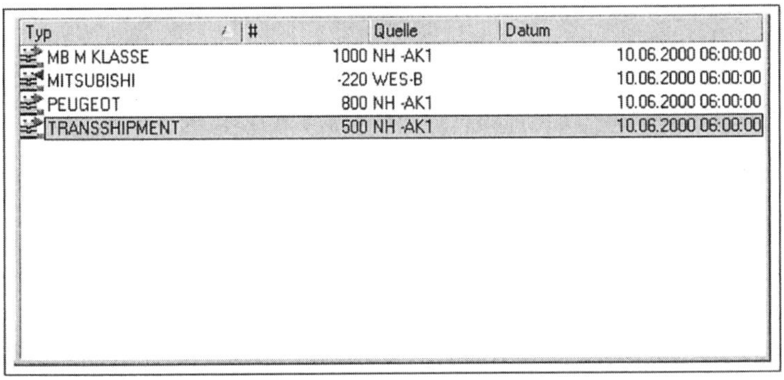
Typ
#
Quelle
Datum
MB M KLASSE 1000 NH -AK1 10.06.2000 06:00:00
MITSUBISHI -220 WES-B 10.06.2000 06:00:00
PEUGEOT 800 NH -AK1 10.06.2000 06:00:00
TRANSSHIPMENT 500 NH -AK1 10.06.2000 06:00:00

*Figure 8.6.* Selected tasks are displayed in the task viewer with respect to storage area "B".

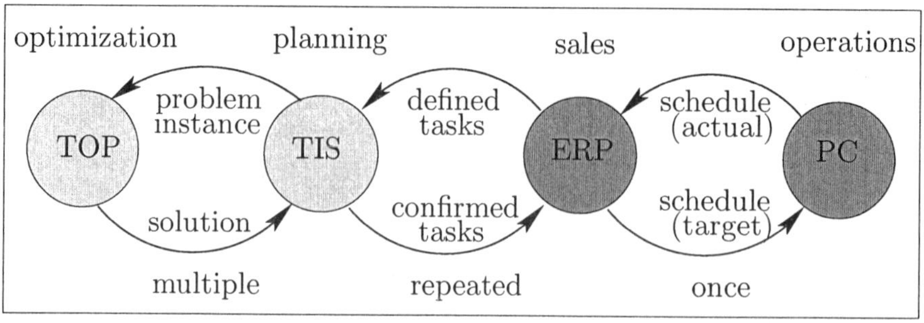


*Figure 8.7.* Business interaction diagram of the planning process.

the following we consider the re-engineering of the process of terminal operations planning. Without automated planning support, the planning and execution of a vehicle transshipment is teamwork of just two organizational units, see the dark-gray shaded systems on the right of Figure 8.7:

1 The sales department defines tasks in the ERP system in accordance with arriving customer orders. After a task has been executed, the corresponding invoice is issued and finally process execution data are provided for controlling purpose.

2 The terminal operation department receives a target schedule from the ERP system and controls the execution of scheduled tasks via PC standard software. The recently finished schedule is transferred back to the ERP system for reasons of accounting.

In case of automated planning support, the planning department is introduced as a new organizational unit. In this extended setting, tasks are no longer scheduled by the sales department. Instead, tasks are processed by TIS and are optimized by TOP.

3 The planning department converts a set of tasks into an optimization problem and transfers this problem to the TOP module. After a solution has been generated, TIS checks this solution for validity and the fully specified tasks are confirmed to the ERP system.

4 The TOP module is evoked as a callable library from TIS. A problem instance is passed, and a solution to the optimization problem is returned. Thereby, TIS performs several consistency checks on the input data and provides an interface for the parameterization of the algorithms of TOP.

The main advantage of this extended planning functionality is the opportunity of frequent re-planning of tasks on the basis of a rolling

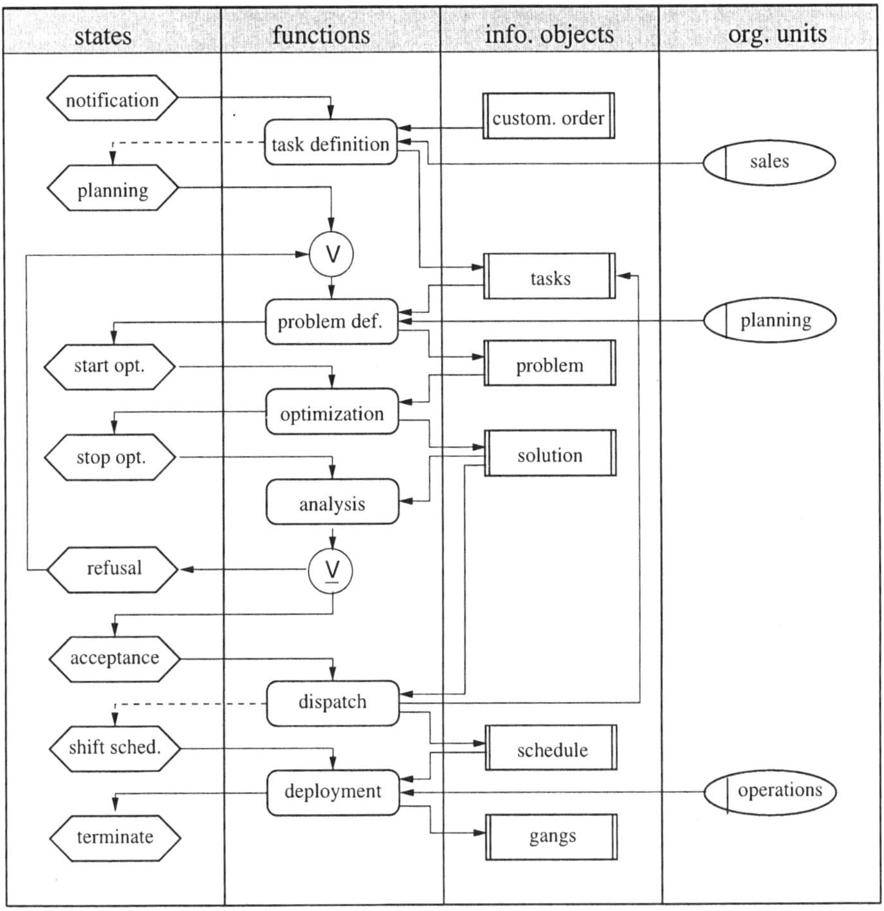


*Figure 8.8.* Process chain diagram of the planning process.

horizon. A task first appears at the end of the planning horizon and is then re-planned for a number of planning iterations before it is eventually confirmed for implementation in the near future.

The process chain diagram of Figure 8.8 illustrates the interplay of the various functions related to planning in more depth. In analogy to the level of requirements definition in ARIS, the diagram shows the interaction of functions and business states in terms of an event driven process chain. Additionally, the diagram integrates the use of information objects and the organizational embedding of functions.

Three organization units are involved in the planning process. The workflow at the interface between different departments is depicted as a dashed line connecting the final function of one subsystem and the initial state of the next subsystem. The sales department receives the customer orders as a notification of dispatch and derives internal tasks from this data. The planning department sets up a problem definition from the partially specified set of tasks and runs an optimization cycle resulting in a solution of the problem.

The solution is carefully analyzed and validated. In case of refusal, the task data are slightly modified and the optimization cycle is repeated. Otherwise the solution is accepted, and a schedule for the forthcoming shift is dispatched to the terminal operations department. The schedule is further processed by a deployment of personnel in terms of the schedule's gang structure, which finishes the planning activities.

Apparently, the information objects generated along the business process resemble each other to a certain extent, cf. the third column from the left hand side of Figure 8.8. First, internal tasks are derived from external customer orders. Then, the subset of tasks with a time windows falling into the current planning horizon are passed on to the problem definition. Apart from the tasks selected and the planning horizon covered, the resources available and the objective pursued also contribute to the constitution of an optimization problem.

For a problem at hand a solution is generated by specifying the storage area, the processing gang and the starting time for every task considered. In refined optimization cycles the problem data will be modified even by means of a re-definition of task data. Once accepted, a solution is dispatched into a schedule, i.e. fully specified tasks for a nearby shift are confirmed for execution. At the same time, decisions concerning later shifts close to the planning horizon are deferred to consecutive planning and scheduling cycles. For the tasks of a schedule a gang structure is specified by means of the personnel deployment.

Since the information objects involved are subsequently used and re-used by the business functions depicted, an integral data model will support the integration of planning and scheduling. The development of such an integral data model is subject to the following considerations.

### 8.3.2 Design Specification

Main purpose of the design specification for business processes is the semi-formal description of data involved. Documentation approaches supporting this goal make use of entity-relationship models (ERM), originally proposed by Chen (1976). An advantage of this modeling approach is that relationships between objects (called entities in this context) are modeled explicitly.

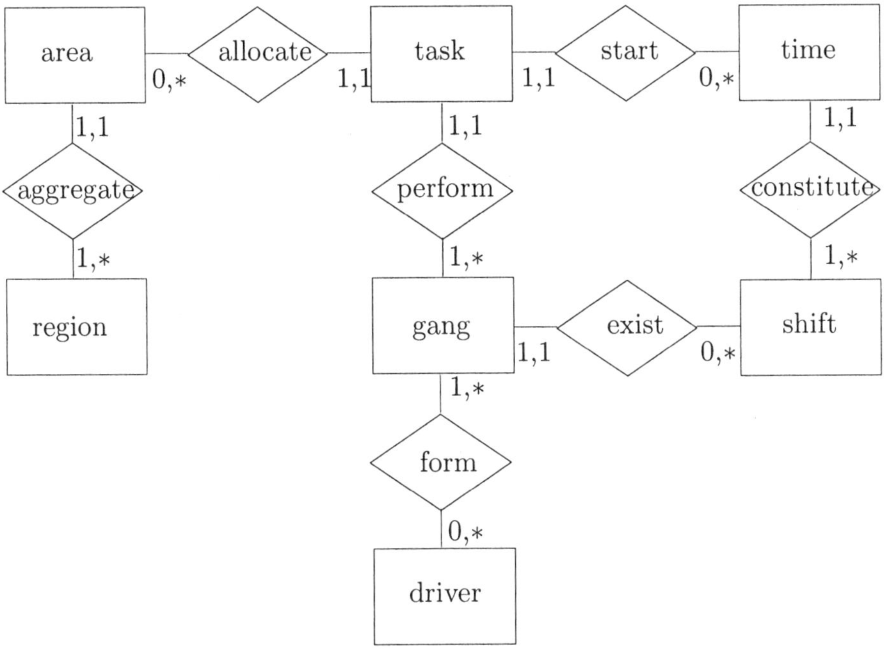


*Figure 8.9.* Objects related to planning in entity-relationship model notation with min/max cardinalities.

Figure 8.9 shows a reasonable, although simplified ERM for data objects related to planning. Rectangular boxes denote entities whereas rhombuses denote relationships. We have chosen an ERM extension, which presents minimum and maximum cardinalities of relationships between entities.

For example, the entities "area" and "task" are coupled by the relationship "allocate". Reading the cardinality from left to right means, that (minimal) zero and (maximal) $n$ tasks allocate a certain storage area. Reading the relationship in the opposite direction means that every task is assigned to exactly one area. Similarly, exactly one gang performs a task, but a gang will typically perform more than one task. A task starts at a discrete time step, and at each time step zero to many tasks can be started.

A prescribed number of time steps constitute a working shift. A gang exists for exactly one shift, although many gangs can exist in a shift. Since the existence of a gang is restricted to one shift, but many shifts are considered, drivers work in many gangs. On the other hand, typically many drivers form a gang. Finally, several areas can be aggregated

*Table 8.1.* We differentiate states of a task with respect to its prescribed attributes and with respect to the attributes acquired by planning. Six meaningful states exist.

| state | prescribed | acquired | system | description |
|---|---|---|---|---|
| UN | undefined | not-known | ERP | task not yet described |
| DN | defined | not-known | ERP | task described, but not yet planned |
| DP | defined | planned | TIS | planning attributes acquired |
| RN | refined | not-known | TIS | prescribed attributes modified |
| RP | refined | planned | TIS | acquired attributes for modified data |
| DS | defined | scheduled | ERP | task fully determined for execution |

into one region, but every area is dedicated to exactly one region. We recognize by means of the (1,1) cardinalities for the task entity in Figure 8.9, that a task is fully specified by a) allocating a certain area, b) a certain gang performing this task and c) the assignment of a unique starting time.

However, little is said about the state transitions necessary for acquiring this data. In Figure 8.8 we identify several information objects as they appear over the course of planning, i.e. a set of tasks, a problem, a solution, a schedule, and finally a gang deployment. The entity-relationship model presented integrates all these information objects. In doing so, the temporal dependencies between the information objects are obscured. A state-transition model can be used to make the dependencies between information objects explicit. States of the transition model can be seen as counterpart to the information objects described on the level of requirements definition.

States of a task with respect to its prescribed attributes (task type, time window, vehicle volume, etc.) can be distinguished. We consider three states,

**undefined** the task is not yet defined in the ERP system,

**defined** the task has been specified by the sales department, and

**refined** the attributes are tentatively changed for planning purposes.

Furthermore we consider the states of task attributes acquired from planning (start time, number of driver, area allocation) . Here, we also distinguish three states,

**non-know** a planning cycle has either not taken place or its results have been discarded,

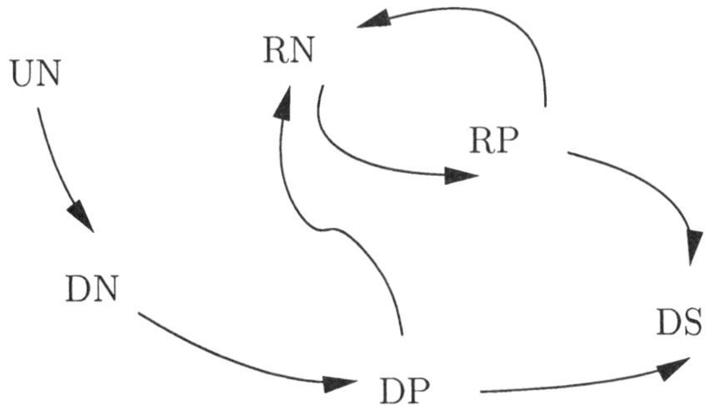


*Figure 8.10.* State transition model for task attributes.

**planned** a planning cycle in the TIS has been successfully performed, and finally

**scheduled** the results of a planning cycle have already been accepted by the ERP system for execution.

Together, the Cartesian product of $3 \times 3 = 9$ states results, but merely six states are meaningful, such that they correspond to an information object, compare Figure 8.8. For example, (DN) corresponds to the information object "problem", because tasks are specified but planning attributes are not yet known. Consequently (DP) corresponds to the information object "solution", because next to the prescribed attributes also the planning attributes have been determined. The six states considered are listed in Table 8.1.

The life cycle of a tasks starts from UN with its definition in the ERP system (DN), see Figure 8.10. The task is considered for planning in the TIS and therefore planning attributes for this task are acquired (DP). The task can be either written back to the ERP system as scheduled for execution (DS), or its prescribed data attributes can be refined in order to generate a different solution in a further planning cycle. In the latter case, its acquired data cannot be used anymore and its prescribed data is noted to be refined (RN). In sub-sequent planning cycles several planning scenarios can be validated (RP). In the event that a refined task is accepted for execution by the ERP system, next to its acquired data also its prescribed data attributes have to be modified in the ERP database.

The formulation of information objects by means of states and state transitions with respect to an integral data model allows the close integration of the planning and scheduling functionality into the existing ERP system. In the remainder of this section we discuss the implemen-

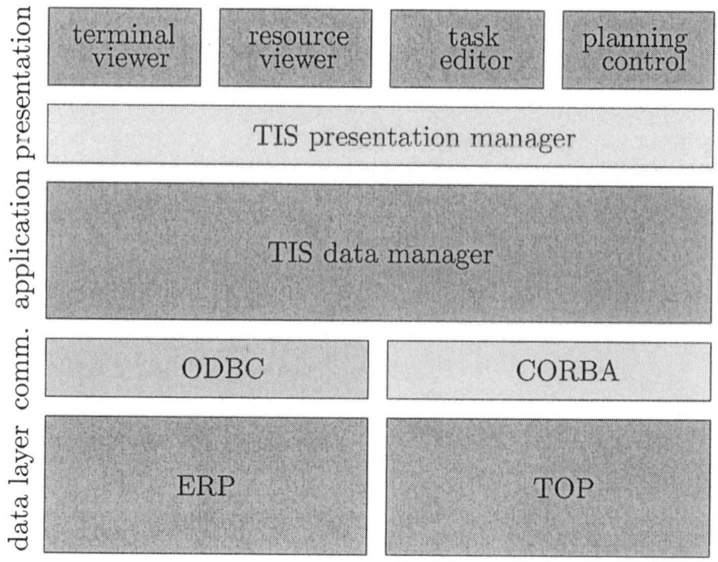


*Figure 8.11.* Model of software layers involved in the Planning solution.

tation of state transition issues in the planning and scheduling software modules.

### 8.3.3 Implementation Description

The IT-integration of planning has been implemented as a distributed system. The ERP system, the TIS and the TOP are autonomous systems and can even run on different hardware platforms. This allows a flexible usage of the system modules, eases maintenance effort significantly, and provides a way to a future re-engineering of individual modules. The modules are separated in accordance to a four-layered model, with the presentation layer further separated into several software components.

Figure 8.11 shows the modules of the entire system arranged in a four-layered structure (Bass et al., 2000). The layers depict the levels of data generation and modification, application and presentation. Additionally, a communication layer is provided in order to physically split the modules of the data layer from the application layer.

The presentation layer consists of several graphical user interface (GUI) components like viewers and editors, compare Section 8.2.2. Furthermore, a planning control window provides a user interface to the TOP module. These GUI components are integrated by means of the TIS presentation manager. As already stated above, its main purpose is the propagation of changing data into the GUI components.

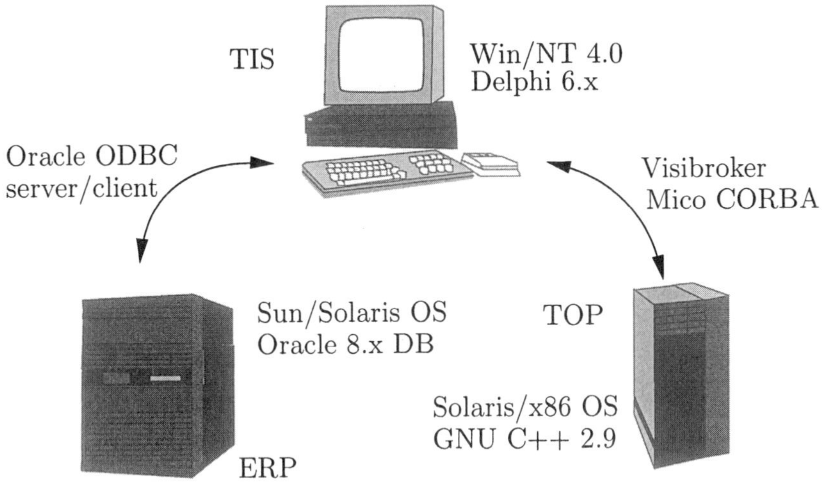


*Figure 8.12.* Hard- and software components involved in planning.

The presentation manager interacts closely with the TIS data manager, which corresponds to the application layer of the four-layered model. The data manager stores an integral model of the planning data and controls its modification by means of a state transition model as described in Section 8.3.2. The data manager is linked via the Open Database Connect Protocol (ODBC) directly to accessible database tables of the ERP system.

Once a planning scenario has been read-in by the application layer and displayed by the presentation layer, the automated planning is performed by the TOP module. This module is connected to the application layer via a Common Open Request Broker Architecture (CORBA) interface, which allows the integration of heterogeneous hard- and software by means of a middle-ware protocol.

The strict appliance of software layers allows the integration of planning in a fairly heterogeneous IT-infrastructure. Figure 8.12 gives a brief overview. The ERP system is based on a SPARC workstation running the Sun/Solaris operating system. The Oracle database software supports remote access via an ODBC server.

The TIS component is developed in Delphi and runs on an ordinary PC under Windows NT. Despite its wide user oriented functionality, its hardware requirement is limited. Obviously, the converse holds for the TOP component. Although heuristics have been chosen to produce a reasonable solution quality in a time span a human planner is willingly to wait for, a fast computer directly increases the response time of planning.

Because of the fast and continuous increase of PC power, a UNIX based operating system has been used on a PC architecture. The TOP module is written in C++ and is interconnected via Mico CORBA to the Visibroker CORBA implementation of the TIS module. Together, all three modules efficiently utilize their distinct hard- and software environments in order to efficiently support the resource planning of the transshipment terminal.

## 8.4 Impact of Automated Planning

The planning and scheduling system described has been in use for the terminal operations of the vehicle terminal in Bremerhaven since January 2001. Now, in 2002 efficiency gains can be reported by comparing productivity measures and transshipment volumes of 2001 and 2002 with the ones of 2000 as the last year of manual planning.

The main challenge of the vehicle terminal in 2001 was to cope with an exceptionally high volume of 1,193 thousand vehicles (in comparison to 1,073 thousand vehicles in 2000). Managing this peak volume by automated planning and scheduling was actually a great success. Generally, an increasing transshipment volume will lead to a decreasing productivity of operations.

Despite the increased transshipment volume marginal productivity gains have been achieved by automated planning and scheduling. The average time of a single vehicle storage took 9.21 minutes in 2000 and was decreased to 9.17 minutes in 2001. With respect to the retrieval of vehicles, 15.2 minutes per unit in 2000 has been reduced to 14.8 minutes in 2001.

After returning to typical load conditions, according to BLG representatives, the productivity i.e. for import transshipment has been increased to currently 16.3% in comparison to 2000. By emanating from 363 employees of regular driving personnel, this figure will lead to an annual reduction of personnel costs of more than 1 million Euro.

Mr. Michael Reiter, the manager of terminal operations, sees the major contribution to this positive development in the process orientation imposed by the automated planning and scheduling system. System modeling and software implementation have changed the managerial focus from inventory management to transshipment processes, such that currently more than 60 % of import vehicles are not relocated beyond the necessary storage and retrieval movements.

Although the system's functioning has surpassed the operator's expectations, further improvements seem possible:

- Along with the development of the planning and scheduling module, the telemetry has been analyzed by the Institute of Shipping Economics and Logistics (ISL). Estimate-actual comparisons of the productivity can further fine-tune the system over time.

- Customers tend not to submit EDI records before the data is entirely definitive. Often this is too late for planning and scheduling purposes; hence the early integration of approximate data from the customer's side is seen as a hallmark for further improvements.

- It's up to the human planner to make use of the system's flexibility. Next to a proper functioning of the user interface, the unswerving belief of the planner that the system will deliver a reasonable solution in every case is of immense importance. Experience will further encourage the planner to entrust planning and scheduling to the automated system.

First reports of the practical use of the system are encouraging. However, the process of organizational embedding of planning has not come to an end yet and therefore further gains can be expected.

## 8.5 Summary

In this chapter we have banked on the problem modeling and heuristics development made in the preceding chapters of this book, by assuming the existence of an efficient terminal operations planning software module. Purpose of this chapter is the possible integration of such a software module into an existing IT-infrastructure of a firm.

By example of vehicle transshipment in Bremerhaven we have described typical functions of ERP systems and their interfaces to the operations execution level. Then, we have outlined requirements to a user interface connecting the ERP system and the automated planning module. Particular attention has been spent on a graphical user interface for the depiction of spatial resources.

Then, we have focused on the planning activities as business process, for which we have proposed a re-engineering step. In analogy to the ARIS toolset we have passed through the requirements definition, where we have defined functions and states of planning. We have assigned information objects to functions and we have embedded those functions into organizational units.

In the design specification we have primarily focused on the data and we developed an integral data model for planning by means of an entity relationship approach. Furthermore, we have identified states for the possible alteration of data attributes. Finally, in the implementation description we have presented a four-layered model of the software modules involved, which allows the physical separation of modules in a distributed computing environment.

# Bibliography


Abrahamsson, M., Brege, S., and Norrman, A. (1998). Distribution channel re-engineering — organizational separation of the distribution and sales functions in the European market. *Transport Logistics*, 1(4):237–249.

Agbegha, G.Y., Ballou, R.H., and Mathur, K. (1998). Optimizing auto-carrier loading. *Transportation Science*, 32(2):174–188.

Amasaka, K. (2002). New JIT: A new management technology principle at Toyota. *International Journal of Production Economics*, 80:135–144.

Angelelli, E. and Speranza, M.G. (2002). The periodic vehicle routing problem with intermediate facilities. *European Journal of Operational Research*, 137:233–247.

Auto & Truck International (2002). World automotive market report 2001-2002. Issued annually by Auto & Truck International, Chicago, Illinois, USA.

Aykin, T. (2000). A comparative evaluation of modeling approaches to the labor shift scheduling problem. *European Journal of Operational Research*, 125:381–397.

Bäck, T. (1996). *Evolutionary Algorithms in Theory and Practice.* Oxford University Press.

Bäck, T., Fogel, D. B., Whitley, D., and Angeline, P.J. (2000). Mutation operators. In Bäck, T., Fogel, D. B., and Michalewicz, T., editors, *Evolutionary Computation 1: Basic Algorithms and Operators*, chapter 32, pages 237–255. IOS Press.

Bahntech (2004). Das Technik Magazin der Deutschen Bahn AG. Brochure by Deutsche Bahn AG, Potsdamer Platz 2, 10785 Berlin, Germany.

Ballou, R.H. (1999). *Business Logistics Management.* Prentice Hall, London, 4. edition.

Bass, L., Clements, P., and Bass, K. (2000). Software architecture documentation in practice. Software Engineering Institute, Carnegie Mellon University, Pittsburgh. Special report of the book "Software Architecture in Practice", Prentice-Hall, 2000.

Bélis-Bergouignan, M.-C., Bordenave, G., and Lung, Y. (2000). Global strategies in the automobile industry. *Regional Studies*, 34(1):41–53.

Bendall, H.B. and Stent, A.F. (2001). A scheduling model for a high speed containership service: A hub and spoke short-sea application. *International Journal of Maritime Economics*, 2(3):262–277.

Beykal, Murat Kemal (2005). Automotive shipping and comparison of planning factors between turkish and world automotive terminals. Master's thesis, Istanbul Technical University, Department of Maritime Transportation Engineering.

Billionnet, A. (1999). Integer programming to schedule a hierarchical workforce with variable demands. *European Journal of Operational Research*, 114:105–114.

Bingle, R., Meindertsma, D., and Oostendorp, W. (1987). Designing the optimal placement of spaces in a parking lot. *Math. Modelling*, 9(10):765–776.

Bish, E.K., Leong, T.-Y., Li, C.-L., Ng, J.W.C., and Simchi-Levi, D. (2001). Analysis of a new vehicle scheduling and location problem. *Naval Research Logistics*, 48:363–385.

Błażewicz, J. and Liu, Z. (1996). Scheduling multiprocessor tasks with chain constraints. *European Journal of Operational Research*, 94:231–241.

Booker, L.B., Fogel, D. B., Whitley, D., Angeline, P.J., and Eiben, A.E. (2000). Recombination. In Bäck, T., Fogel, D. B., and Michalewicz, T., editors, *Evolutionary Computation 1: Basic Algorithms and Operators*, chapter 33, pages 256–269. IOS Press.

Borstnar, C. (1999). *The Positioning of South Korean Companies in Europe.* PhD thesis, Universität St. Gallen. No. 2264.

Böse, J., Reiners, T., Steenken, D., and Voß, S. (2000). Vehicle dispatching at seaport container terminals using evolutionary algorithms. In Sprague, R. H., editor, *Proceedings of the 33rd Annual Hawaii International Conference on System Sciences*, pages 1–10. IEEE.

Böttcher, J., Drexl, A., Kolisch, R., and Salewski, F. (1999). Project scheduling under partially renewable resource constraints. *Management Science*, 45:543–559.

Bramel, J. and Simchi-Levi, D. (1997). *The Logic of Logistics*. Springer Series in Operations Research. Springer.

Bremenports (2003). bremenports GmbH & Co.KG, Masterplan zur Optimierung des Automobile-Logistics-Centers Bremerhaven. Market Analysis.

Brucker, P., Drexl, A., Möhring, R., Neumann, K., and Pesch, E. (1999). Resource-constrained project scheduling: Notation, classification, models, and methods. *European Journal of Operational Research*, 112:3–41.

Brusco, M.J. and Johns, T.R. (1996). A sequential integer programming method for discontinuous labor tour scheduling. *European Journal of Operational Research*, 95:537–548.

Bubois, A. and Gadde, L.-E. (1997). Information technology and distribution strategy. In Tilanus, B., editor, *Information systems in logistics and transportation*, pages 33–56. Pergamon.

Carlier, J. (1982). The one-machine scheduling problem. *European Journal of Operational Research*, 11:42–47.

Cassady, C.R. and Kobza, J.E. (1998). A probabilistic approach to evaluate strategies for selecting a parking place. *Transportation Science*, 32(1):30–42.

Chen, P.P.-S. (1976). The entity-relationship model — toward a unified view of data. *ACM Transactions on Database Systems*, 1(1):9–36.

Chen, T. (1999). Yard operations in the container terminal — a study in the 'unproductive moves'. *Maritime Policy and Management*, 26(1):27–38.

Cohen, M.A. and Mallik, S. (1997). Global supply chains: Research and applications. *Production and Operations Management*, 6(3):193–210.

Cullen, T.J. (1998). *European Finished Vehicle Logistics.* Cargo Systems Ltd., IIR Publications, 5th Floor, 29 Bressenden Place, London SW1E 5DR.

Daganzo, F.D. (1999). *Logistic Systems Analysis.* Springer.

Daniels, R.L. (1990). A multi-objective approach to resource allocation in single machine scheduling. *European Journal of Operational Research*, 48:226–241.

Dantzig, G.B. (1954). A comment on edies's traffic delays at toll booths. *Operations Research*, 2:339–341.

Deb, K. (2000). Introduction to selection. In Bäck, T., Fogel, D. B., and Michalewicz, T., editors, *Evolutionary Computation 1: Basic Algorithms and Operators*, chapter 22, pages 166–171. IOS Press.

Dodin, B., Elimam, A.A., and Rolland, E. (1998). Tabu search in audit scheduling. *European Journal of Operational Research*, 106(373–392).

Domschke, W. and Krispin, G. (1997). Location and layout planning: A survey. *Operations Research Spektrum*, 19:181–194.

Dornier, P.-P., Ernst, R., Fender, M., and Kouvelis, P. (1998). *Global Operations and Logistics.* John Wiley & Sons, New York.

Dreo, Johann, Petrowski, Alain, Siarry, Patrick, and Taillard, Eric (2005). *Metaheuristic Optimization : Methods and Case Studies.* Springer, Berlin.

Drewry (1999). *Market Outlook for Car Carriers.* Drewry Shipping Consultants Ltd., Drewry House, Meridian Gate – Sourth Quay, 23 Marsh Wall, London E14 9FJ, England.

Drozdowski, M. (1996). Scheduling multiprocessor tasks — an overview. *European Journal of Operational Research*, 94:215–230.

Ebben, M.J.R., Heijden, M.C. van der, and Harten, A. van (2004). Dynamic transport scheduling under multiple resource constraints. *European Journal of Operational Research*, 167:320–335.

Fagerholt, K. (2000). Evaluating the trade-off between the level of customer service and transportation costs in a ship scheduling problem. *Maritim policy and management*, 27(2):145–153.

Fagerholt, K. and Christiansen, M. (1999). A combined ship scheduling and allocation problem. Technical report, Dept. of Marine Systems

Design, Norvegian University of Science and Technology, Trondheim, Norway.

Feitelson, D. G. (1996). Packing schemes for gang scheduling. In Feitelson, D. G. and Rudolph, L., editors, *Job Scheduling Strategies for Parallel Processing*, volume 1162 of *Lecture Notes in Computer Science*, pages 89–110. Springer Verlag.

Fink, Andreas, Voß, Stefan, and Woodruff, David L. (1999). Intelligent heuristic search conponentware. TU Braunschweig, Schriftenreihe des Inst. f. Wirtschaftswissenschaften.

Fischer, T. and Gehring, H. (2004). Planning vehicle transshipment in a seaport automobile terminal using a multi-agent system. *European Journal of Operational Research*, 166:726–740.

Fischer, Torsten (2003). *Multi-Agenten-Systeme im Fahrzeugumschlag.* PhD thesis, Fern-Universität Hagen, Germany.

Fleischmann, B. (2005). Distribution and transport planning. In Stadtler, H. and Kilger, C., editors, *Supply Chain Management and Advanced Planning*, pages 229–244. Springer.

Galliers, R.D. (1994). Information systems, operational research and business reengineering. *International Transaction on Operational Research*, 1(2).

Garstone, Sue (1995). Electronic data interchange (EDI) in port operations. *Logistics Information Management*, 8(2):30–33.

Gavish, B. and Pirkul, H. (1991). Algorithms for the mulit-resource generalized assignment problem. *Management Science*, 37(6):695–713.

Giannopoulos, G.A. (2004). The application of information and communication technologies in transport. *European Journal of Operational Research*, 152:302–320.

Glover, F. (1989). Tabu search–part i. *ORSA Journal on Computing*, 1:190–206.

Glover, F. (1990). Tabu search–part ii. *ORSA Journal on Computing*, 2:4–32.

Glover, F. and Laguna, M. (1993). Tabu search. In Reeves, C. R., editor, *Modern Heuristic Techniques for Combinatorial Problems*, pages 70–150. Blackwell.

Glover, Fred and Kochenberger, Gary A., editors (2003). *Handbook of metaheuristics.* Kluwer, Boston.

Glover, Fred and Laguna, Manuel (2002). *Tabu Search.* Kluwer, Boston, 6th edition.

Gnoni, M. G., Iavagnilio, R., Mossa, G., Mummolo, G., and Leva, A. Di (2003). Production planning of a multi-site manufacturing system by hybrid modelling: A case study from the automotive industry. *International Journal of Production Economics*, 85(2):251–262.

Goetschalckx, M., Vidal, C.J., and Dogan, K. (2002). Modeling and design of global logistics systems: A review of integrated strategic and tactical models and design algorithms. *European Journal of Operational Research*, 143:1–18.

Goetshcalckx, Marc and Ratliff, H. Donald (1990). Shared storage policies based on the duration stay of unit loads. *Management Science*, 36(9):1120–1132.

Goldberg, D.E. (1989). *Genetic Algorithms.* Addison-Wesley.

Goldratt, E.M. (1997). *Critical Chain.* North River Press, Great Barrington.

Graves, G.W., McBride, R.D., Gershkoff, I., Anderson, D., and Mahidhara, D. (1993). Flight crew scheduling. *Management Science*, 39(6):736–745.

Günther, Hans-Otto and Kim, K.-H. (2004). Container terminal logistics. OR Specturm special issue 1-2, volume 26,.

Gupta, Y.P. and Somers, T.M. (1992). The measurement of manufacturing flexibility. *European Journal of Operational Research*, 60:166–182.

Haase, K. (1993). *Lotsizing and Scheduling for Production Planning*, volume 408 of *Lecture Notes in Economics and Mathematical Systems.* Springer Verlag.

Hafenvertretung, Bremische (2003). Jahresbericht. Electronic version can be obtained from http://www.bhv-bremen.de.

Hall, R.W. (1987). Comparison of strategies for routing shipments through transportation terminals. *Transportation Research A*, 21:421–429.

Harman, R. (2000). Second tier ports in danger from industry consolidation. *Automotive Logistics*, July/September, pages 63–66.

Hartmann, S. (2004). A general framework for scheduling equipment and manpower at container terminals. *OR Spectrum*, 26:51–74.

Herfort, R. (2002a). Handling with care. *Automotive Logistics,* July/September, pages 38–47.

Herfort, R. (2002b). Maritime Fahrzeuglogistik: Hafenwahl in Europa. Deutsche Verkehrszeitung, Sonderbeilage Automobillogistik, 29.10.2002.

Hertz, A., Taillard, E., and de Werra, D. (1997). Tabu search. In Aarts, E. H. L. and Lenstra, J. K., editors, *Local Search in Combinatorial Optimization*, chapter 5, pages 121–136. Wiley.

Hesse, M. (2002). Shipping news: the implications of electronic commerce for logistics and freight transport. *Resources, Conservation and Recycling*, 36:211–240.

Hines, P., Silvi, R., and Bartolini, M. (2002). Demand chain management: an integrative approach in automotive retailing. *Journal of Operations Management*, 20:707–728.

Holguìn-Veras, Josè and Jara-Dìaz, Sergio (1999). Optimal pricing for priority service and space allocation in container ports. *Transportation Research B*, 33:81–106.

Holland, H.J. (1975). *Adaptation in natural and artificial systems.* The University of Michigan Press, Ann Abor.

Holocher, K.H. (2000). Marktstrukturen des Automobilumschlags in den Häfen der Nordrange. Market Analysis.

Holweg, M., Disney, S. M., Hines, P., and Naim, M. M. (2005). Towards responsive vehicle supply: A simulation-based investigation inot automotive scheduling systems. *Journal of Operations Management*, 23(5):507–530.

Holweg, Matthias and Miemczyk, Joe (2003). Delivering the '3-day car' — the strategic implications for automotive logistics operations. *Journal of Purchasing and Supply Management*, 9(2):63–71.

Iranpour, R. and Tung, D. (1989). Methodology for optimal design of a parking lot. *Journal of Transportation Engineering*, 115(2):139–160.

Jiang, B., Huang, B., and Vasek, V. (2003). Geovisualisation for planning support systems. In Geertman, S. and Stillwell, J., editors, *Planning Support Systems in Practice*, Advances in spatial science, pages 177–192. Springer, Berlin.

Jones, C. and Clark, J. (1990). Effectiveness framework for supply chain management. *Computer Integrated Manufacturing Systems*, 3(4):196–206.

Keijzer, M., Merelo, J.J., Romero, G., and M.Schoenauer (2001). Evolvable Objects: A general purpose Evolutionary Computation library. In *5th Conference Internationale sur l' Evolution Artificielle.* Universite de Bourgogne, France. Web: http://eodev.sourceforge.net.

Kia, M., Shayan, E., and Ghotb, F. (2000). The importance of information technology in port terminal operations. *International Journal of Physical Distribution & Logistics Management*, 30(3/4):331–344.

Kiedel, Klaus-Peter (2001). Autotransport. Brochure of the German Maritime Museum. Electronic version can be obtained from http://www.dsm.de/auto.

Kim, Kap Hwan and Park, Kang Tae (2003). A note on a dynamic space-allocation method for outbound containers. *European Journal of Operational Research*, 148:92–101.

Kim, Ki Young and Kim, Kap Hwan (1999). A routing algorithm for a single straddle carrier to load export containers onto a containership. *International Journal of Production Economics*, 59:425–433.

King, J. (1997). Globalization of logistics management: present status and prospects. *Maritim policy and management*, 24(4):381–387.

Kobrin, S.J. (1991). An empirical analysis of the determinants of global integration. *Strategic Management Journal*, 12:17–31.

Kolisch, R. (1995). *Project Scheduling under Resource Constraints.* Physica Verlag, Heidelberg.

Kozan, Erhan and Preston, Peter (1999). Genetic algorithms to schedule containers transfers at multimodal terminals. *Intl. Trans. in Op. Res.*, 6:311–329.

Kuhn, J. (1998). Der Global Player Ansatz im Automobilbau. *Zeitschrift für Betriebswirtschaft*, 68(9):937–957.

Kuhr, M. (2000). Extending the hub-and-spoke concept in the vehicle trades. In *Global Automotive Trades & Logistics Conference and Exhibition.* Lloyd's List Events.

Kusiak, A. (2000). *Computational Intelligence in Design and Manufacture.* Wiley.

Kwan, R.S.K. and Wren, A. (1996). Hybrid genetic algorithms for bus driver scheduling. In Bianco, I. and Toth, P., editors, *Advance Methods in Transportation analysis*, pages 609–619. Springer Verlag.

Laborie, P. (2001). Algorithms for propagating resource constraints in AI planning and scheduling: Existing approaches and new results. In Cesta, A. and Borrajo, D., editors, *Proceedings of the Sixth European Conference on Planning*, pages 205–216.

Laguna, M., Kelly, J.P., González-Velarde, J.L., and Glover, F. (1995). Tabu search for the multilevel generalized assignment problem. *European Journal of Operational Research*, 82:176–189.

Leeper, J.H. (1988). Integrated automated terminal operations. *Transportation Research Circular*, 33(2):23–28.

Lieske, T. (2002). Marktanteils- und Portfolioanalyse nach Marktsegmenten im Fahrzeugumschlag der Wettbewerbshäfen und Umschlagbetriebe in der Hamburg — Antwerpen / Zeebrügge Range. Master's thesis, Universität Bremen, Lehrstuhl für Logistik.

Lim, A. (1998). The berth planning problem. *Operations research letters*, 22:105–110.

Lüders, C. (2001). Die Entwicklung des Automobilmarktes ASEAN. Bremer Lagerhaus Gesellschaft, Market Analysis.

Mangan, J., Lalwani, C., and Gardner, B. (2002). Modelling port/ferry choice in RoRo freight transportation. *International Journal of Transport Management*, 1:15–28.

MarketLine (1998). *EU Automotive Logistics*. MarketLine International Ltd., 16 Connaught Street, London, W22AF, England.

Marle, G. van (2003). The whether forecast. *Automotive Logistics*, January/March, pages 33–38.

Mason, S.J., Ribera, P.M., Farris, J.A., and Kirk, R.G. (2003). Integrating the warehousing and transportation functions of the supply chain. *Transportation Research Part E*, 39(141–159).

Mattfeld, D.C. (1996). *Evolutionary Search and the Job Shop*. Physica Verlag, Heidelberg.

Mattfeld, D.C. and Kopfer, H. (2003). Terminal operations management in vehicle transshipment. *Transportation Research A*, 37:435–452.

Mattfeld, Dirk C. (2003). Mid-term planning of transshipment tasks. In Spengler, T., Voß, S., and Kopfer, H., editors, *Logistikmanagement 2003*, pages 411–424. Springer.

Mattfeld, Dirk C. (2005). It-integration of terminal operations planning. In Günther, H.-O., Mattfeld, D. C., and Suhl, L., editors, *Supply Chain Management und Logistik*, pages 319–336. Physica.

Mattfeld, Dirk C. and Branke, Jürgen (2005). Task scheduling under gang constraints. In Kendall, G., Burke, E., Petrovic, S., and Gendreau, M., editors, *Multidisciplinary Scheduling; Theory and Applications*, pages 113–130. Springer.

Mayer, G. (2001). *Strategische Logistikplanung von Hub&Spoke-Systemen.* Deutscher Universitäts-Verlag, Wiesbaden.

Mehrotra, A., Murphy, K.E., and Trick, M.A. (2000). Optimal shift scheduling: A branch-and-price approach. *Naval Research Logistics Quarterly*, 47:188–200.

Michalewicz, Z. (1996). *Genetic Algorithms + Data Structures = Evolution Programs.* Springer, 3rd edition.

Miemczyk, J. and Holweg, M. (2001). Build-to-order: Rethinking the automotive supply chain. In *Survey on Vehicle Logistics 2001*, pages 273–278. European Car-Transport Group of Interest, Madrid, Spain.

Mourão, M.C., Pato, M.V., and Paixão, A.C. (2001). Ship assignment with hub and spoke constraints. *Maritime Policy and Management*, 29:135–150.

Muralidharan, B., Linn, R.J., and Pandit, R. (1995). Shuffling heuristics for the storage location assignment in an AS/RS. *International Journal on Production Research*, 33(6):1661–1672.

Murty, K.G., Liu, J., Wan, Y., and Linn, R. (2005). A decision support system for operations in a container terminal. *Decision Support Systems*, 39(3):309–332.

Narasimhan, R. (1996). An algorithm for single shift scheduling of hierarchical workforce. *European Journal of Operational Research*, 96:113–121.

Neumann, K. and Schwindt, C. (1999). Project scheduling with inventory constraints. Technical Report WIOR-572, University of Karlsruhe.

Ng, J.K.C., Ip, W.H., and Lee, T.C. (1999). A paradigm for ERP and BRP integration. *International Journal of Production Research*, 37(9):2093–2108.

Nijkamp, P., Pepping, G., and Banister, D. (1996). *Telematics and transport behaviour*. Advances in spatial science. Springer, Berlin.

Nijkamp, Peter and Ubbels, Barry (2004). Infrastructure, suprastructure and ecostructure: a portfolio of sustainable growth potentials. Vrije Universiteit Amsterdam, Serie Research memoranda, Faculteit der Economische Wetenschappen en Econometrie.

Nishimura, E., Imai, A., and Papadimitriou, S. (2001). Berth allocation planning in the public berth system by genetic algorithms. *European Journal of Operational Research*, 131:282–292.

Nishimura, E., Imai, A., and Papadimitriou, S. (2005). Yard trailer routing at a maritime container terminal. *Transportation Research E*, 41(1):53–76.

Nobert, Y. and Roy, J. (1998). Freight handling personnel scheduling at air cargo terminals. *Transportation Science*, 32(2):295–301.

O'Kelly, M.E. (1986). The location of interacting hub facilities. *Transportation Science*, 20:92–106.

Paixão, A.C. and Marlow, P.B. (2003). Fourth generation ports — a question of agility? *International Journal of Physical Distribution & Logistics Management*, 33(4):355–376.

Petersen, Charles G. (1999). The impact of routing and storage policies on warehouse efficiency. *International Journal of Operations & Production Management*, 19(10):1053–1064.

Phillips, L.T. (1987). Air carrier activity at major hub airports and changign interline practices in the united states' airline industry. *Transportation Research A*, 21:215–221.

Polewa, R., Lumsden, K., and Sjöstedt, L. (1997). Information as a value adder for the transport user. In Tilanus, B., editor, *Information systems in logistics and transportation*, pages 157–168. Pergamon.

Pontrandolfo, P. and Okogbaa, O.G. (1999). Global manufacturing: a review and a framework for planning in a global corporation. *International Journal of Production Research*, 37(1):1–19.

Porter, M.E. (1980). *Competitive strategy: techniques for analyzing industries and competitors.* Free Press, New York.

Preston, Peter and Kozan, Erhan (2001). An approach to determine storage locations of containers at seaport terminals. *Computers & Operations Research*, 28:983–995.

Racunica, I. and Wynter, L. (2000). Optimal location of intermodal freight hubs. Technical report, INRIA Research Report 4088, Institut National de Recherche en Informatique et en Automatique, Cedex, France.

Reeves, C.R. (1993). Genetic algorithms. In Reeves, C. R., editor, *Modern Heuristic Techniques for Combinatorial Problems*, chapter 4, pages 151–196. Blackwell.

Rizzi, A. and Zamboni, R. (1999). Efficiency improvement in manual warehouses through ERP systems implementation and redesign of the logistics processes. *Logistics Information Management*, 12(5):367–377.

Rodrigue, J.P. (1999). Globalization and the synchronization of transport terminals. *Journal of Transport Geography*, 7:255–261.

Ronen, D. (1993). Ship scheduling: The last decade. *European Journal of Operational Research*, 71:325–333.

Sagan, Hans and Bishir, John W. (1991). Optimal allocation of storage space. *European Journal of Operational Research*, 55:82–90.

Salewski, F., Schirmer, A., and Drexl, A. (1997). Project scheduling under resource and mode identity constraints: Model, complexity, methods, and application. *European Journal of Operational Research*, 102:88–110.

Scheer, A.-W. (1999). *ARIS - business process modeling.* Springer, Berlin, 3rd edition.

Schmitz, J. and Platts, K. W. (2004). Supplier logistics performance measurement: Indications from a study in the automotive industry. *International Journal of Production Economics*, 89(2):231–243.

Schneeweiss, C. (1999). *Hierarchies in Distributed Decision Making.* Springer.

Schneeweiss, C. and Zimmer, K. (2004). Hierarchical coordination mechanisms within the supply chain. *European Journal of Operational Research*, 153:687–703.

Schwefel, Hans-Paul (1977). *Numerische Optimierung von Computer-Modellen mittels der Evolutionsstrategie.* Birkhäuser.

Schwindt, Christoph (2002). Introduction to resource allocation problems in project management. *Habilitation thesis*, University of Karlsruhe.

Shabayek, A.A. and Yeung, W.W. (2002). A simulation model for the Kwai Chung container terminals in Hong Kong. *European Journal of Operational Research*, 140:1–11.

Shobrys, D.E. and White, D.C. (2002). Planning, scheduling and control systems: why cannot they work together. *Computers and Chemical Engineering*, 26:149–160.

Smith, J.M. (1989). *Evolutionary Genetics.* Oxford University Press, Oxford.

Snyder, R.L. and Ibrahim, H.A. (1996). Capacity planning and operational assessment of bulk storage systems. *Production and Inventory Management Journal*, 37:14–18.

Spatz, J. and Nunnenkamp, P. (2002). Globalization of the automobile industry — traditional locations under pressure? *Aussenwirtschaft*, 57(4):469–493.

Steenken, D., Henning, A., Freigang, S., and Voß, S. (1993). Routing of straddle carriers at a container terminal with the special aspect of internal moves. *OR-Spektrum*, 15:167–172.

Steenken, Dirk, Voß, Stefan, and Stahlbock, Robert (2004). Container terminal operation and operations research. *OR Spectrum*, 26(1):3–49.

Supply-Chain Council (2002). Supply chain operations reference model — Version 5.0. Technical report, Pittsburgh.

Sürie, C. and Wagner, M. (2005). Supply chain analysis. In Stadler, H. and Kilger, C., editors, *Supply Chain Management and Advanced Planning*, pages 37–63. Springer Verlag, 3rd edition edition.

Taleb-Ibrahimi, M. and Castilho, B. de (1993). Storage space vs handling work in container terminals. *Transportation Research B*, 27(1):13–32.

Thompson, R.G. and Richardson, A.J. (1998). A parking search model. *Transportation Research A*, 32(3):159–170.

Tulder, R. van and Ruigrok, W. (1998). International production networks in the auto industry: Central and Eastern Europe as the low end of the West European car complexes. In Zysman, J. and Schwarz, A., editors, *Enlarging Europe: The Industrial Foundations of a new Political Reality*, no. 99, pages 202–237. University of California at Berkeley.

Tyan, J.C., Wang, F.-K., and Du, T.C. (2003). An evaluation of freight consolidation policies in global third party logistics. *Omega*, 31:55–62.

Verband der Automobilindustrie (2002). Auto Jahresbericht 2002. Issued annually by VDA, Frankfurt am Main, Germany.

Vis, I.F.A. and de Koster, R. (2003). Transshipment of containers at a container terminal: An overview. *European Journal of Operational Research*, 147:1–16.

Wan, T.B., Wah, E.L.C., and Meng, L.C. (1992). The use of IT by the port of Singapore Authority. *World Development*, 20(12):1785–1795.

Ward's Communication (var.). WARD'S automotive yearbook. Issued annually by Ward's Communication, Southfield, Minnesota, USA.

Wiendahl, H.-P. (1987). *Belastungsorientierte Fertigungssteuerung.* Hanser Verlag, München.

Williams, H.P. (1999). *Model Building in Mathematical Programming.* Wiley, 4 edition.

Woodbridge, C. (2001). Challenges ahead for car carriers. *Seatrade Review,* May.

Woodruff, David L. and Voß, Stefan, editors (2002). *Optimization software class libraries.* Kluwer Acadademic Publisher, Boston.

Yun, W.Y. and Choi, Y.S. (1999). A simulation model for container-terminal operation analysis using an object-oriented approach. *International Journal of Production Economics*, 59:221–230.

Zhang, C., Liu, J., Wan, Y., Murty, K.G., and Linn, R.J. (2003). Storage space allocation in container terminals. *Transportation Research Part B*, 37:883–903.

# Index